"Raymo is not ambivalent about where he stands. As a skeptic, he champions a brutal honesty with respect to beliefs, allowing them only something approaching hypothetical status. He is indebted to the rigor of scientific method. He also finds in skepticism a resilient and expansive outlook, and he combines this with an openness to the awe and wonder of the scientifically disclosed universe and life within it." —*Science*

"Responding in part to the rise of millennial-driven New Age spirituality, Raymo writes along the tender edges of mystery that bind off objective science from religious faith. Using a light journalistic style, Raymo seeks to find some common ground upon which to construct mutual appreciation between science and religion. Sources diverse as John Donne, Charles Darwin, Gerard Manley Hopkins and Albert Einstein enliven the discussion. . . . A scientist through and through, Raymo yet maintains an appreciation for the ineffable in life." —*Publishers Weekly*

"A provocative and important book on a tough range of subjects. . . . [Raymo's] bravery in tackling such an elusive subject is in line with the best scientific minds and their unquenchable desire to know and understand more. . . . He also has a fantastic ability to describe the miracles and mysteries that science has shown us in nature. . . . As you read [*Skeptics and True Believers*], you'll be forced to think and rethink through comfortable assumptions concerning science, religion, and the nature of the relationship between the two." —*Astronomy*

"What sets this book apart is the fact that Raymo respects both the scientific need to understand and the religious need to celebrate creation." —*Choice*

"The hard truth Raymo so judiciously presents is that we must get over our fear and loathing of science so that we can create a new set of beliefs that unites the revelations of both the tangible world and the soul." —*Booklist*

"Raymo argues that religion should embrace the reliable knowledge of the world that science provides, while at the same time science should respect and nourish humankind's need for spiritual sustenance." —*Reference & Research Book News*

"Contemporary science has provided us with a new and fascinating creation story. But how does it square with the Book of Genesis and traditional religious faith? This is one question that Raymo takes up in this dazzling book on the current debate between science and religion. Raymo is a witty and clear-headed guide through the thickets of current science—from evolutionary biology to the miracles of DNA codes and theories of the Big Bang. At the same time he is a sensitive interpreter of authentic religious faith and its compatibility with the scientific search." —David Toolan, S.J.

Skeptics and True Believers

THE EXHILARATING CONNECTION BETWEEN SCIENCE AND RELIGION

Chet Raymo

WALKER AND COMPANY
NEW YORK

First published in the United States of America in 1998 by Walker Publishing Company, Inc.; first paperback edition published in 1999.

Library of Congress Cataloging-in-Publication Data
Raymo, Chet.
Skeptics and true believers: the exhilarating connection between science and religion/Chet Raymo.
p. cm.
Includes bibliographical references and index.
ISBN 0-8027-1338-6 (hardcover)
1. Religion and science. I. Title.
BL240.2.R36 1998
215—dc21 98-14647
CIP

ISBN 0-8027-7564-0 (paperback)

Book design by Mauna Eichner

Printed in the United States of America

2 4 6 8 10 9 7 5 3

To Maureen

To invoke God as a blanket explanation of the unexplained is to make God the friend of ignorance. If God is to be found, it must surely be through what we discover about the world, not what we fail to discover.

Paul Davies, physicist

When it's over, I want to say: all my life
I was a bride married to amazement.
I was the bridegroom, taking the world into my arms.

Mary Oliver, poet

Contents

Skeptics and True Believers

Introduction

THERE'S A "God-shaped hole in many people's lives," says physicist and Anglican priest John Polkinghorne. He's right, at least about there being a hole in our lives. To call the hole "God-shaped" begs the question, for the affliction of our times is that we have no satisfactory image of God that rests comfortably with what scientists have learned about creation. As we approach the end of the twentieth century, many educated people in the Western world long wistfully for something akin to traditional religious faith, but they know there can be no turning back to a world of divine fiats and penny miracles. As Polkinghorne says, they can neither accept the idea of God nor quite leave it alone.[1]

I am one of those people, trained in science, who cannot quite accept the idea of God nor quite leave it alone. I am less pessimistic than most, however, that science and religion must remain in conflict. It seems to me that science is part of the traditional religious quest for the God of creation.

A vital religious faith has three components: a shared cosmology (a story of the universe and our place in it), spirituality (personal response to the mystery of the world), and liturgy (public expressions of awe and gratitude, including rites of passage). The apparent antagonism of science and religion centers mostly on cosmological questions: What is the universe? Where did it come from? Where is it going? What is the human self? Where do we fit in? What is our fate?

Humans have always had answers to these questions. The answers have been embodied in stories: tribal myths, scriptures, church traditions. All of these stories have been derived from a primordial experience of the creation. All of them contain enduring wisdom. But for many of us, these stories have been superseded *as public knowledge* by the scientific story of the universe.

In this book, I identify two intellectual postures we can adopt to questions of knowledge and faith. These two postures represent a fault line in our culture, an attitudinal chasm more profound than differences of politics or religious affiliation.

We are Skeptics or True Believers.

Skeptics are children of the Scientific Revolution and the Enlightenment. They are always a little lost in the vastness of the cosmos, but they trust the ability of the human mind to make sense of the world. They accept the evolving nature of truth and are willing to live with a measure of uncertainty. Their world is colored in shades of gray. They tend to be socially optimistic, creative, and confident of progress. Since they hold their truths tentatively, Skeptics are tolerant of cultural and religious diversity. They are more interested in refining their own views than in proselytizing others. If they are theists, they wrestle with their God in a continuing struggle of faith. They are often plagued by personal doubts and prone to depression.

True Believers are less confident that humans can sort things out for themselves. They look for help from outside—from God, spirits, or extraterrestrials. Their world is black and white. They seek simple and certain truths, provided by a source that is more reliable than the human mind. True Believers prefer a universe proportioned to the human scale. They are repulsed by diversity, comforted by dogma, and respectful of authority. True Believers go out of their way to offer (sometimes forcibly administer) their truths to others, convinced of the righteousness of their cause. They are likely to be "born again," redeemed by faith, apocalyptic. Although generally pessimistic about the state of *this* world, they are confident that something better lies beyond the grave.

In the following pages, I will closely examine the

cultural implications of these two frames of mind, specifically with regard to science and religion. Not all religious people fit into the category defined here as True Believers, just as not all scientists are Skeptics. If a Skeptic is one who is willing to live with a measure of doubt, then Job, Jesus ("My God, my God, why hast thou forsaken me?"), Pascal, Graham Greene, Dietrich Bonhoeffer, Martin Buber, and many other great religious leaders, writers, and thinkers have been Skeptics. On the other hand, those scientists who are invincibly certain of the authority of their science must be counted as True Believers.

Yet science, as we understand it today, can only thrive among Skeptics. Some constancy of belief is essential for any way of knowing, but science is by definition driven by research, open to growth and even, on occasion, to revolutionary change. Einstein once remarked that the most important tool of the scientist is the wastebasket. A scientist must be skeptical of her most cherished theory; if she is not, then others within the scientific community will do it for her. Indeed, science is little more than organized skepticism. A successful scientific idea must run a fierce gauntlet of peer review. Our confidence in scientific knowledge is based on trial by fire: systematized doubt and persuasion. Generally, the competition of scientific ideas acts—as in biological evolution—to refine the status quo, sharpening the match between theory and perceptions. Sometimes, however, the perceptual landscape of science radically shifts—as with the invention of the microscope or telescope, or the discovery of electromagnetic waves—and new ideas suddenly emerge and triumph.

True Believers have low tolerance for changeable knowledge. They prefer stable truths of faith, even if those truths run counter to a preponderance of physical evidence. For example, a 1993 Gallup poll indicates that nearly half of Americans believe in the idea of a geologically young Earth, despite the fact that not a shred of reproducible empirical evidence can be adduced in favor of the idea and a mountain of evidence is arrayed against it.

The forces that nudge us toward True Belief are pervasive and well-nigh irresistible. Supernatural faith systems provide a degree of emotional security that skepticism cannot provide. Who among us would *not* prefer to believe that there exists a divine parent who has our best interest at heart? Who among us would *not* prefer to believe that we will live forever? Skepticism, on the other hand, offers only uncertainty and doubt. What keeps scientific skepticism on track, against the individual's need for emotional security, is a highly evolved social structure, including professional associations and university departments, peer-reviewed literature, meetings and conferences, and a language that relies heavily on mathematics and specialized nomenclature. The point of this elaborate apparatus is to minimize individual backsliding into the false security of True Belief. Political, cultural, linguistic, and religious idiosyncrasies are suppressed in favor of the common endeavor.

Of course, a danger remains that the scientific community might lapse into True Belief—for example, that the theory of adaptation of organisms by natural selection might become unbreachable dogma. But the history of science suggests that even the most thoroughly entrenched ideas (absolute space-time or fixed continents, for example) can be made untenable by a recalcitrant mismatch of theory and observation. In a 1996 story in *Time* magazine, John Bahcall, a physicist at the Institute for Advanced Study in Princeton, New Jersey, expressed confidence in the reliability of current theoretical models for what happens at the center of the Sun; then he added, "But that's why you do experiments. Because what you think you know might turn out to be completely wrong."

For a dozen years, I have written a weekly column about science and nature for the *Boston Globe.* In it, I have explored the ways that scientific knowledge impinges upon our personal and public lives. During those dozen years, no topic has evoked more reader response than the intersection of science and faith. I have received correspondence on this topic from hundreds of readers, the overwhelming majority of whom offered thoughtful, provocative, and helpful responses. One theme emerged from this exchange of ideas: We are a culture divided at its heart. We warmly embrace the technological and medical fruits of science, but often hold religious beliefs that stand in flat-out contradiction to the scientific way of knowing. We concede that science has proved spectacularly successful as a way of understanding the world, yet firmly reject one of its clearest implications: *We are ephemeral, contingent parts of a silent universe that is vastly larger than ourselves.*

Some of us respond to this apparently deflating concept of who we are by embracing faiths that emphasize our personal cosmic importance—fundamentalist religions, New Age superstitions, pseudosciences—all patently at odds with empirical knowledge. Others seek a place for the doctrines of traditional faith in the gaps of science, in the supposed uncertainties of quantum physics, or in the mathematics of chaos. Still others talk about "complementary ways of knowing," and so compartmentalize their minds that they are able to keep science and traditional religious doctrines from coming into conflict.

As for myself, I was raised in a traditional Christian faith. I took academic degrees in science and found in science a compelling cosmological vision of the world. I did not turn science into a religion; science is too shallow a vessel to hold ultimate mysteries. At the same time, nothing I had been taught in my religious education seemed adequate to encompass the grandeur and mystery of what I learned in science. It became obvious to me that certain doctrines of the Judeo-Christian tradition, including such central tenets of faith as immortality and a personal God who answers prayers, were based on long-discredited views of the world that placed humans in a central position and ascribed human attributes to other creatures and even to inanimate objects. At the same time, I found within the Judeo-Christian tradition vital mystical and liturgical practices that nourished my quest for encounter with the Absolute.

I am today a thoroughgoing Skeptic who believes that words like *God*, *soul*, *sacred*, *spirituality*, *sacrament*, and *grace* can retain currency in an age of science, once we strip them of outworn overlays of anthropomorphic and animistic meaning. Like many others in today's society, I hunger for a faith that is open to the new cosmology—skeptical, empirical, ecumenical, and ecological—without sacrificing historical vernaculars of spirituality and liturgical expression.

Science, too, can only gain from a reconciliation of science and faith. Science sometimes is aloof, arrogant, blind to the ambient mystery that animates knowledge. A fusion of knowledge with religious feeling need not weaken the rigor of scientific skepticism; it can, however, help stitch science back into the larger fabric of our emotional, intuitive, aesthetic, and sensual lives.

ONE

Miracles and Explanations

See! I am God. See! I am in everything. See! I never lift my hands off my works, nor will I ever. See! I lead everything toward the purpose for which I ordained it.

Julian of Norwich

LIKE MOST CHILDREN, I was raised on miracles. Cows that jump over the moon; a jolly fat man that visits every house in the world in a single night; mice and ducks that talk; little engines that huff and puff and say, "I think I can"; geese that lay golden eggs. This lively exercise of credulity on the part of children is good practice for what follows—for believing the miracle stories of traditional religion, yes, but also for the practice of poetry or science.

Science is based upon our ability to imagine what we cannot see: nuclear reactions in the cores of stars, the spinning of galaxies, the dervish dance of DNA. Science, like the imaginative landscapes of childhood, is a world of make-believe. It is, however, a very special kind of make-believe. Science takes as given that a real world exists "out there," and that it can be represented, albeit imperfectly, in the world of ideas. We struggle mightily to make the partition between the imagined world and the real world as transparent as possible. No scientist will dispute that "atom" is a made-up concept; however, the concept "atom" is the *most concise* way—perhaps the only way—to make sense of our detailed, quantitative experience of the material world. Without the concept "atom," chemistry, X-ray crystallography, nuclear energy, thermodynamics, and other broad territories of external experience make no sense at all. Indeed, so transparent is the partition between "atom" and experience that most scientists would say that atoms are "facts," or at least so close to being facts that no quotation marks are called for.

In the Land of Make-Believe

It is because we retain as adults something of the child's facility for make-believe that we can enthuse with the poet Gerard Manley Hopkins:

Look at the stars! Look, look up at the skies!
O look at all the fire-folk sitting in the air![1]

It is also because we retain something of the child's facility for make-believe that we can imagine that the stars are vast spheres of hydrogen and helium, powered by nuclear energy, light-years away. Poetic metaphor ("fire-folk") and scientific construct (nuclear-powered spheres of gas) serve useful functions in our lives, but we are confident the latter bears a closer affinity to reality—to whatever is "out there"—than the former. The poetic metaphor conveys a human truth; the scientific construct attempts to remove the human subject from the equation of idea and reality.

The biologist Richard Dawkins has suggested that the credulity of children—the willingness to believe whatever one is told by adults, especially parents—has been reinforced by natural selection for its survival value.[2] The child comes into the world knowing nothing, and must quickly learn how to navigate the perils of life. At first, "Don't touch the stove" and "Be good or Santa won't bring toys" are absorbed with equal credulity. The child is asked by an authority figure to behave as if the stove is hot, and to behave as if Santa exists, and so she does. The challenge of growing up is to learn which sorts of make-believe are useful reality constructs and which are poetic metaphors.

Early on in our lives, we abandon Santa Claus and the tooth fairy as reality constructs because we recognize contradictions that are difficult to resolve (the relative sizes of Santa's rotund belly and the chimney pipe, for example), but also because word gets around from other presumably reliable authorities, older siblings perhaps, that the stories are untrue. As for the stove, we learn to exercise a certain skepticism concerning whether or not it is hot, testing in doubtful cases by cautiously touching the surface with a fingertip.

We cannot live without some sorts of make-believe in our lives. Without made-up maps of the world, life is a blooming, buzzing confusion. Some elements of our mental maps (Santa Claus, fire-folk) satisfy emotional or aesthetic *inner needs*; other elements of our mental maps (hot stove, nuclear-powered stars) satisfy intellectual curiosity about the world *out there*. We get in trouble when the two kinds of maps are confused, when we objectify elements of make-believe solely on the basis of inner need. No one takes *fire-folk* literally; but many of us accept the astrological influence of the stars on our lives because it satisfies an inner need, even in the face of convincing evidence to the contrary (every objective test of astrology has proved negative).

The True Believer retains in adulthood an absolute faith in some forms of empirically unverifiable make-believe (such as astrology or the existence of immortal souls), whereas the Skeptic keeps a wary eye even on firmly established facts (such as atoms). Both Skeptic and True Believer use made-up maps of the world.

Is one map as good as any other? Since all knowledge is constructed, can the choice between two contradictory maps (fire-folk versus nuclear-powered spheres of gas, for example) be a matter of personal or political expediency? Not unless we are willing to erect partitions between what we believe to be true on the basis of unambiguous, reproducible evidence and what we merely wish to be true. Apparently, many of us are willing to do just that. A 1995 Gallup poll showed that 79 percent of adult Americans believe in miracles (interestingly, 86 percent of women believe in miracles, compared to 71 percent of men). About half of us are open to the reality of astrological influences. Nearly three-quarters of us believe in life after death. When teenagers were asked, "When scientific and religious explanations conflict, which explanation are you more likely to accept?" the majority chose religion by a factor of two to one.

The Unmiraculous Shroud

A linen cloth preserved in the cathedral at Turin, Italy, the Shroud of Turin, bears the likeness of a man and is purported to be the winding sheet of Christ. The cloth has long been an object of veneration among Christians. In the late 1980s, Roman Catholic authorities allowed scientists to take tiny samples of the shroud for radiocarbon dating. This technique uses the precisely known decay rate of radioactive carbon atoms as a kind of clock to determine when organic substances—bone, wood, charcoal, et cetera—were alive.[3] The method has enjoyed wide use among archaeologists, paleontologists, and historians. It has been calibrated against the ring count of ancient trees and tested successfully many times on historical objects of known age.

In the case of the Shroud of Turin, carbon dating shows when the flax plants were alive from which the linen was made. Three independent carbon-dating labs, in Zurich, Oxford, and Tucson, Arizona, participated in the test.[4] Along with a sample from the shroud, each lab was given three control samples of cloth of known age: linen from a 900-year-old Nubian tomb, linen from a second-century mummy of Cleopatra, and threads from an 800-year-old garment of St. Louis d'Anjou. None of the samples was identified for the researchers. None of the labs communicated with the others until the results were in. After making their measurements, all three labs agreed on the ages of all four samples. All three labs correctly dated the control samples. And all three labs concluded that the Shroud of Turin is medieval, dating from the mid–fourteenth century. Significantly, this is the very time the shroud first appears in historical records.

It is to the credit of Church officials in Italy that they authorized the carbon-dating tests and accepted the results. Their actions are in keeping with a declaration by Pope John Paul II on the relationship of science and theology: "Science can purify religion from error and superstition, and religion can purify science from idolatry and false absolutes."[5]

Is the conclusion of the radiocarbon tests absolute? No, of course not. No scientific test can prove anything with absolute certainty. Is the conclusion convincing? Yes, if you are a Skeptic. No, if you are a True Believer. The person with True Belief in the shroud's authenticity will dismiss any evidence to the contrary.

In fact, carbon dating of the Turin shroud seems only to have enhanced its reputation as the winding sheet of Christ. (Web pages on the Internet are devoted to its cult.) Since the test results were announced, many attempts have been made to explain them away. According to one critic, a burst of neutrons from the body of the risen Christ created extra carbon-14 nuclei, making the cloth appear younger than it actually is. (No mention is made of what might have caused this mysterious neutron burst, other than a miracle.) Another critic has suggested that the presence of bacteria on the cloth might have muddied the result by adding modern-day carbon-14, although no evidence is adduced that such bacteria actually exist on the Shroud of Turin. As I write, several Italian professors claim to have seen the image of a first-century Roman coin on the cloth. No test, no matter how carefully contrived, will dissuade a True Believer from his belief. Given a conflict between scientific and religious explanations, most of us are quite willing to go with the religious explanation if it confirms our deep-seated inner need for miracles.

Early in my education, the Shroud of Turin was offered to me as evidence for the risen Christ, and therefore for the truth of Christianity. I was educated in Roman Catholic schools, where miracles were as much a part of the curriculum as Dick and Jane and the multiplication tables. *The Shroud of Turin. The spinning Sun at Fátima. Having our throats blessed with crossed candles on Candlemas Day, thereby making us immune to choking on chicken bones. St. Brendan the Navigator taking refuge on the back of a whale during his sixth-century voyage to America (my teachers were Irish nuns). Et cetera, et cetera.* We lived within a vast and engaging landscape of miracles, as richly improbable (by empirical standards) as the make-believe landscape of fairy tales, and including, of course, those constant miracles we had with us every day: the Real Presence of Christ's body and blood in the Eucharist, the efficacy of intercessory prayer, angels, devils, heaven, limbo, purgatory, hell, and life everlasting. I absorbed these things, mostly uncritically, because it is the nature of children to be credulous. I didn't ask for evidence; the miracles *were* the evidence.

In retrospect, it is easy to see that the entire panoply of miracles, including the most outrageously improbable—all those little unbaptized babies in limbo, for instance—were there to bolster the possibility that death is not final. St. Paul said (as we were frequently reminded), "If the dead are not raised, Christ has not been raised, and if Christ has not been raised, your faith is in vain." (1 Cor. 15:17) The Shroud of Turin and all the rest were offered as evidence that our ultimate fate is not to be food for worms.

By the time I went off to the University of Notre Dame, many of the more fanciful miracles of my primary education had faded from the story, but the big miracles remained. The text we used for my freshman theology class was Frank Sheed's *Theology and Sanity*, the thrust of which was that any sane person *must* be a Roman Catholic, so persuasive is the evidence for the objective truth of that faith. Meanwhile, I was studying science and discovering a way of constructing mental maps of the world that allowed no place for miracles.

This is not to say that science proves miracles are impossible. One does not prove the invalidity of a miracle by showing that it is inconsistent with the laws of nature. It is the nature of miracles—the strength of their force as evidence—that they violate natural law. Science works by finding consistent patterns in nature; miracles, if they occur, are by definition one-time things. In my university science classes, I did not learn that miracles are impossible, but that there is no reliable evidence that they occur.

Every miracle, examined closely, has a way of slipping through the fingers. En masse the evidence for miracles looks impressive; but take them one at a time and they become frustratingly evasive. As I searched among the miracles of my faith, I found none that was not contaminated with the likelihood of flawed testimony, fraud, or wishful thinking. Always there was the possibility of a natural explanation. The person whose illness abates after a trip to Lourdes might have been cured by the intercession of the Virgin, but the illness also might have receded on its own or have been ameliorated by positive thinking; both circumstances are recognized within the natural order. The Shroud of Turin might be the winding sheet of the risen Christ, but it might also be an ingeniously contrived fraud or work of art, both of which were common in the fourteenth century. Shine the fierce light of skepticism on the Lourdes cure or the Turin shroud and the "miracle" vanishes.

The Miraculous Red Knot

I learned something else in my study of science, something that had an even greater effect upon my religious faith: None of the miracles I had been offered in my religious training were as impressively revealing of God's power as the facts I was learning in science. In one of his sermons, the poet John Donne writes: "There is nothing that God hath established in a constant course of nature, and which therefore is done every day, but would seem a miracle, and exercise our admiration, if it were done but once."[6] Consider, for example, the flight of juvenile red knots from the islands of northern Canada to Tierra del Fuego, at the southern tip of South America.

The red knot is a sandpiper that twice each year visits the eastern shores of the United States. Every year, these tough little travelers wing more than 18,000 miles, from the southern tip of South America to the arctic islands of northern Canada and back again, stopping briefly along the way on the beaches of Delaware Bay and Cape Cod.

During our northern winter, red knots feed on the sunny beaches of Tierra del Fuego. The birds take advantage of the austral summer to replace their tattered feathers in a long molt, which ensures their flight equipment is in top condition when, in

February, they lift off in flocks of hundreds or thousands for the journey north. Up the coast of Argentina, across the hump of Brazil, stopping occasionally along the way to fatten up. They know exactly where to find food, returning each year to the same stretches of sand or marsh. From the northern coast of South America, they strike out across the Atlantic on a weeklong nonstop flight that brings them in mid-May to their usual feeding grounds on the marshy shore of Delaware Bay, just as horseshoe crabs are laying eggs by the millions.

For a few weeks the red knots gorge themselves; a single bird might consume 135,000 horseshoe crab eggs. Then, fat and fit again, they take to the air for a nonstop flight to islands of the Canadian archipelago north of Hudson Bay. Here, in the boreal summer, they mate and breed, each female red knot laying four speckled eggs, which she and her mate incubate in turns. Baby knots are up and about as soon as they hatch, growing rapidly and replacing natal down with juvenile feathers in preparation for flight. By mid-July, the female adults abandon their new offspring and head south; male adults follow a few weeks later. The juveniles fend for themselves until late August, when they too commence the 9,000-mile journey to Tierra del Fuego.[7]

Now here is the astonishing thing, and the reason I have told the story. The young red knots, by the thousands and *without adult guides or prior experience,* find their way along the ancient migration route. From northern Canada to New England's Atlantic shore, across the Atlantic Ocean to Guyana and Suriname, then down along the eastern coast of South America, arriving precisely at those feeding grounds along the way where they are sure to find food. At last they join their parents and others of their species on the beaches of Tierra del Fuego for the southern summer.

How do they do it? How do the young birds make their way along a route they have never traveled to a destination they have never seen? How do they unerringly navigate the long stretch of their journey over featureless sea? We know exactly *what* the red knots accomplish—where they go, when they arrive; dedicated ornithologists have banded the birds by the hundreds, watched for them at way stations, counted their comings and goings. But *how* the uninstructed young birds accomplish their epic feat of navigation remains mysterious. The Sun, the stars, the Earth's magnetic field, angles of polarized light—all of these have been shown to be part of the navigational skills of one animal or another (birds, fish, or insects), and singly or in combination these clues must keep the red knots on course.

This much is certain: A map for the journey and the instrumental knowledge to follow it are part of the red knot's genetic inheritance. Each bird begins life as a single fertilized cell. Already, that microscopic cell contains the biological equivalent of a set of charts, a compass, a sextant, and maybe even something akin to a satellite navigation system. This must be true, for every bird is born with the instinct to make its journey.

How can a map of the globe and the skill to follow it be contained within a cell too small to be seen with the naked eye? Medieval theologians are said to have debated how many angels can dance on the head of a pin; in the flight of the red knot we are engaged with a mystery more immediately present but no less marvelous. We can call it instinct and let it go at that. But human curiosity will not let it go. We ask: How? The need to find answers is deep within us, anchored at the root of our being. Of all species of life on Earth, we are the one that *wants to know*. We want knowledge that is reliable, public, and universal, based upon unambiguous, reproducible experience that is (or can be) common to all of us—in a word, knowledge that is scientific.

In the case of animal navigation, the answer to our question turns out to be quite incredible. The urge to make the red knot's planet-spanning flight, the map of the journey, and the skills to follow it, are written into a DNA molecule in a language of stunning simplicity. The molecule is shaped like a spiral staircase—the famous double helix. The side rails of the staircase are linked sugar and phosphate molecules. The treads are paired molecules called nucleotides. There are four kinds of nucleotides: adenine, guanine, cytosine, and thymine, designated A, G, C, and T. Adenine always pairs with thymine, and guanine always pairs with cytosine, so that there are four kinds of treads along the DNA staircase: A-T, T-A, G-C, and C-G. It is the sequence of these treads that is the genetic code. The red knot's map and navigational manual are written in a chemical language of only four letters!

In each cell of the red knot's body, there are identical strands of DNA, about an arm's length in all, a blueprint for making a small russet bird with an urge to fly and the skills to make a 9,000-mile unpracticed journey. Can it be possible? There are thirty-two volumes in the *Encyclopaedia Britannica,* 1,000 pages per volume, 1,200 words per page, an average of five letters per word, for a total of 200 million letters. There are several billion nucleotide pairs in an arm's length of DNA. A sequence of three nucleotide pairs (sixty-four possible combinations) is enough to provide a code for each letter of the alphabet, upper- and lowercase plus punctuation. Believe it or not, several sets of the *Encyclopaedia Britannica* could be transcribed into the red knot's genes!

This information is not in doubt. Molecular biologists can isolate DNA, replicate it, photograph it, measure it, read the sequence of nucleotides, change the sequence, modify genes. It is possible in principle to provide a complete transcription of the red knot's

DNA (this has been done for many organisms), and to determine which parts of the sequence code for eye, feather, beak, claw. Somewhere along the red knot's double helix—somewhere among those many volumes of information—is the code for constructing those parts of the red knot's brain that contain the map of the migration route and the skills to follow it. The red knot's brain is a flexible organ, capable of wiring itself by experience. But part of the red knot's brain comes already wired with a map of the globe and a navigator's skills.

For some years I have been on the Board of Overseers of Boston's Museum of Science. On my visits to the museum, I always make my way to the ten-foot-high model of a segment of DNA. To my mind, it is the most extraordinary exhibit in the museum. Atoms are represented by colored balls—carbon black, oxygen red, nitrogen blue, hydrogen white—linked by rods. The model contains only a few dozen pairs of nucleotides, a tiny fraction of what is contained within the DNA of even the simplest living organism. If the whole of the red knot's complement of DNA were shown at the scale of the model, it wouldn't fit within the entire museum. Nevertheless, I stand in front of this partial strand, gape-jawed at the beauty, at the simplicity—a simplicity out of which emerges the astonishing diversity and awesome complexity of life. What I feel as I stand before the model cannot be adequately put into words. Call it reverence, awe, praise—in short, the full range of religious feeling.

Nothing I learned during my religious training is more wondrous to me than the flight of the juvenile red knot from northern Canada to Tierra del Fuego, a journey whose map is contained in the red knot's DNA. Such real-world mysteries inspire my awe far more than the so-called miracle on display in the cathedral at Turin. In the red knot's story, we catch a glimpse of a God who never lifts his hand from his work, and who leads everything to the purpose for which it was ordained. As the British writer and cartographer Tim Robinson observed: Miracles are explainable; it is the explanations that are miraculous.

TWO

Decoding the Mystery of Life

This is the land of lost content,
I see it shining plain,
The happy highways where I went
And cannot come again.

A. E. Housman

THE DIFFERENCE BETWEEN Skeptics and True Believers is not that Skeptics believe what is sensible and obvious, while True Believers accept what is fanciful and far-fetched. Often, it is the other way around. Scientific concepts can be extraordinarily bizarre, as strange to our notion of what is a proper landscape as are the mountainous frozen oceans and sulfur fountains of Jupiter's moons. By contrast, the True Believer's world has a touching familiarity—the kindly, gray-bearded God who reaches out to touch Adam's fingertip on the ceiling of the Sistine Chapel, for example, or those pop-eyed little extraterrestrials who step from flying saucers. On the face of it, scientific concepts would seem to require a far greater degree of credulity than the anthropomorphic projections of the True Believer.

Some time ago, I saw in the journal *Science* a photograph of bacterial DNA, made with an electron microscope. If you have ever seen the mess a kitten makes of a ball of twine, this is it. The DNA in the photograph was twisted into knots and loops as intricate as a crocheted tablecloth. The author of the accompanying article used the metaphor "elaborate fishnet." Then he went on to say, "Yet somehow the fishnet manages to reproduce itself."[1]

Somehow indeed!

It has always seemed to me a miracle that even a short linear strand of DNA might reproduce itself. In principle the idea is simple enough. When the time comes to reproduce, the double-helix "staircase" unzips down the middle. Each unzipped half-staircase turns itself into a full staircase by taking chemical components from the surrounding medium—new bits of banister and tread. The idea is stunningly ingenious, perhaps the most important scientific discovery of the twentieth century. Every high school biology student absorbs the lesson: DNA unzips; each half-strand acts as a template for building a full strand; a copy of the replicated DNA moves to each side of the cell; the cell divides. One cell makes two. Two make four. Four make eight. *Et cetera, et cetera.* Zip, split. Zip, split. Life goes on.

But things are not quite as simple as the textbook makes it seem. If you stretched out the DNA in a single human cell, it would reach from fingertip to fingertip of your outstretched arms. In the trillions of cells in the human body, there is enough DNA, if stretched out, to reach to the Sun and back a dozen times! The DNA in a cell is tangled into forty-six chromosomes. Replication starts at hundreds or thousands of sites, at precisely defined moments in the cell's reproductive cycle. Billions of chemical units in the DNA must be copied exactly, exactly once, no more, no less. Any foul-up can be damaging or fatal to the organism.

The genetic material is active not only during cell division; it is writhing and twisting all the time, zipping and unzipping here and there along the strands, generating fernlike traceries as it spins off RNA and builds proteins—a whirling-dervish dance of life. Look again at that electron microphotograph of the tangled strand of DNA. It is not just lying there, static, like a cast-aside string of pearls. It is a pulsing, undulating farrago of threads, feathers, knobs, and whiskers, a microscopic lace maker frenetically making a lace called life.

That this should happen, minute by minute, hour by hour, in every cell of our bodies, without resulting in a hopeless tangle is—to put it bluntly—unbelievable. The more we learn of the details, the more miraculous it seems. No matter how much we read about *how* it happens, no matter how many electron microphotographs we see of the actual process in its various stages, no matter how hard we try to stretch our imaginations to encompass the DNA's fandango dance of life, it seems impossible.

And yet I believe it.

But I don't believe in the gray-bearded God of the Sistine Chapel ceiling, extraterrestrial visitors, or the Shroud of Turin as Christ's winding sheet, all of which, on the face of it, are more plausible.

For part of each year, I live in rural Ireland on a hillside track known locally as "The Fairies' Road." Once, my neighbor expressed a reluctance to walk the road at night, apprehensive, even in this age of science, of the little folk under the hill. I was not able to suppress a condescending smile. Later I mentioned to this same person that our hill had once been covered with an Ice Age glacier. Now it was her turn to smile. "It is easier to believe in fairies under the hill," she said, "than ice on top."

And of course she's right.

Fairies, Dragonflies, and DNA

So why do I believe in the unerring fandango dance of the DNA, which I cannot fully imagine no matter how hard I try, and not fairies, which any child can imagine? To answer this question adequately would take a book, but to provide a brief answer I'll borrow a diagrammatic scheme proposed by physicist-philosopher Henry Margeneau.[2]

Down the middle of a blank page Margeneau draws a vertical line that he calls the "perception plane." It represents the locus of our immediate sensations of the world—sights, tastes, odors, touches, sounds—the interface between the world *as it is* and the world *as we know it.* To the left of the line is the world "out there," which we know only through the windows of our senses. To the right of the line Margeneau draws circles representing "constructs": names, descriptions, or ideas we invent to make sense of our perceptions. The more abstract the construct, the farther he places the circle from the line.

Immediately adjacent to the perception plane are those constructs that correspond to direct sensations: "blue," "bitter," "pungent," "brittle," "shrill." The construct "dragonfly" is a bit farther from the perception plane, but not very far away. I feel a sensation on my finger ("tingle"), I see a color ("blue"), a quality of light ("iridescent"), a shape ("long and narrow"), and I name this ensemble of sensations "dragonfly." Perhaps I am the first person to see a dragonfly, in which case I invent the name; I have added a circle to my conceptual map of the world ("dragonfly"), linked by short sturdy lines to other circles more closely adjacent to the perception plane ("tingle," "iridescent blue," "long and narrow"). Or perhaps I recognize a congruence between the current ensemble of sensations and other sensations stored in memory; I have seen a dragonfly before, or a picture of a dragonfly. In which case, the remembered construct is reinforced by my new experience; the links between the circles are made bolder, darker; the construct is anchored yet more firmly to the world "out there." I close my eyes and open them again; the links hold. I move my finger; the sensations change; the form flutters, moves away, I hear a *whirr*, the tingle ceases.

Perception and cognition are hugely complex processes, endlessly debated by psychologists, neurologists, and philosophers. Margeneau's simple schematic of connected circles is itself only a construct, a useful way of describing the devilishly complex business of perception and cognition. The important thing is to realize that our ideas about the world are not the same as the world itself (a point often missed by True Believers). Nevertheless, only the most obtuse idealist would hesitate to call "dragonflies" real.

In the case of dragonflies, at least, we are confident that the construct captures the reality.

What about the construct "fairy"? We should also put "fairy" rather close to the perception plane. After all, there is nothing terribly abstract about fairies; a fairy is a little person with dragonfly wings, and there is nothing unfamiliar about any of that. Perhaps our ancestors invented the construct "fairy" because of some primitive intuition that we are not alone. Whatever the reason, once we have the construct, it is easy enough to make connections. Let's say a tool goes missing from the garden, or we hear a strange singing noise on the hill at night; between the construct "fairy" and "missing tool" and "singing noise" we draw lines, anchoring "fairy" into our map of reality. What is missing from our map are lines connecting the construct "fairy" directly to immediate sensations. No one has actually seen a fairy.

"DNA replication" is a construct far removed from the perception plane. There is almost nothing about the construct that relates to ordinary experience, which is why the construct is so difficult to imagine. The perceptions upon which the construct is based are highly technical—for example, X-ray diffraction photographs and demanding chemical assays. The construct "double-helix DNA" is connected with reality by way of many other technical constructs, circles connected to circles in a vast web, by as many paths as we can devise and test, until at last we reach the relevant immediate perceptions—blackened grains in a photographic emulsion, for example, or a reading on a microbalance in the chemical lab—perceptions that mean nothing except in the context of the entire web of constructs. The scientist looks for taut and unambiguous connections between constructs and perceptions that can be subjected to quantitative and reproducible experimental tests.

You will have noticed that I have used familiar metaphorical language in describing the activity of DNA: *fandango dance, stairs, banisters, treads, zipper,* and so on. It should be clear, however, that this language *in itself* does not anchor the construct "double-helix DNA" firmly to reality. We do not suppose that the connection between a DNA molecule and a zipper is anything more than metaphorical. By contrast, to speak of a fairy as a "little person with dragonfly wings" is the essence of the construct.

If we accept Margeneau's schematic, and admit that "fairy" and "double-helix DNA" are both mental constructs removed from immediate sensation, then why are we allowed to say that one construct is more or less "real" than the other? Should we consider "fairy" to be more real since it is closer to the perception plane and therefore more familiar? The True Believer has heard the singing on the hill. She has experienced the tool gone missing from the garden. Therefore fairies exist and that's the end of it. The Skeptic would say, however, that the links connecting "fairy" to the noise on the hill and to the missing tool are flimsy. The noise might just as well be explained by the construct "wind," with links that are firmer, more reproducible, and more widely acknowledged. The missing tool might be attributed to absentmindedness or human theft, both of which are universally acknowledged parts of our common experience. In other words, "fairy" is connected to immediate sensation by few and arbitrary lines. Snip away the construct "fairy" and the rest of the map stands firm, no sensations go unexplained.

Our understanding of DNA replication, on the other hand, is embedded in a vast and resonant web of interconnected constructs. It is the essence of scientific skepticism to test and retest each link in the web, to try to prove it faulty, to look for more concise patterns of constructs and connections that will adequately explain our immediate sensations—the blackened grains in the photographic emulsion, the results of the chemical assay. If we have succeeded in constructing a resonant web of constructs, then any observer, Skeptic or True Believer, should be able to trace the links back to the perceptional source along vibrant lines of connection. It is the firmness of these many connections, based upon tens of thousands of *exact, quantitative, reproducible* experiments, that anchors the construct "double-helix DNA" to reality. Snip a line of connection here and there—the web still holds. Remove the construct entirely, and sensations go unexplained. And *that's* why we believe in the seemingly impossible dance of the DNA.

But it isn't easy. Many of the links in the scientist's map of the world are highly technical. Only narrow specialists will comprehend some of the connections. Any one scientist must trust the veracity of all other scientists, which is why so much effort goes into quantitative data keeping, citation of relevant prior research, and peer review. A scientist giving a talk to fellow scientists, even to close colleagues, is unlikely to get very far before someone interrupts with "Now wait a minute, about that last step . . ." I have often watched the skeptical engine of science at work—winnowing, pruning, testing the resilience of the web. You don't want to be on the receiving end of this kind of collective scrutiny unless your ducks are well in line.

Certainly it is easier to believe in fairies than in DNA. It is also more consoling, more self-edifying, more entertaining. Fairies play into the whole gamut of human emotions: love, fear, power, powerlessness, the "land of lost content" of childhood. But "fairies" are a concept we can do without and still make perfect sense of the world. We cannot do without the concept "DNA" whether we are Skeptics or True Believers.

Metaphors for Life

Once in my *Boston Globe* column I mentioned that an arm's length of DNA resides in every cell of the human body. A correspondent wrote to say that no matter how hard she tried, she could not imagine an arm's length of anything curled up in a cell that is microscopically small. I thought about it. She is right. It *is* damnably difficult. Here was something I had accepted without question for more than forty years because I had read it in books, and when it comes right down to it, it sounds impossible. So I did a calculation.

We know from X-ray diffraction studies that the DNA double helix is about three nanometers in diameter (a nanometer is a billionth of a meter). A typical animal cell is fifteen micrometers in diameter (a micrometer is a millionth of a meter). For purposes of calculation, I represented the DNA as a threadlike cylinder about a meter long and the cell as a sphere. It turns out, having done the calculation, that an arm's length of DNA is hundreds of times less voluminous than a cell. What seemed impossible now makes sense. Moral: Mathematics can be an indispensable aid to the imagination.

A child's imagination is trained on concepts that reside close to the perception plane: parents, teddy bear, security blanket, house, tree. But move away from the world of immediate perception, into the universe of the galaxies and the DNA, and our imaginations falter. Familiar metaphors are found wanting. So we scrape around and invent new ones. Recall what happens in every cell of our bodies as DNA replicates itself: Replication starts at hundreds or thousands of sites, at precisely defined moments in the cell's reproductive cycle; billions of chemical units in the DNA strand are copied exactly, exactly once; any mistake can be damaging or fatal; this happens unceasingly throughout our lives. How? How can it possibly be true? Where can we find an aid for our imagination?

I can think of only one thing in our common experience that begins to approach the complexity and finesse of DNA replication: computers. On a microcomputer chip, tens or hundreds of millions of invisibly small electronic switches open and shut each second, in precise sequence, with essentially zero tolerance for error. If even one of the millions of switches fails to operate at the correct moment, the system crashes. Yet computer chips run reliably for days or years on end without any visible manifestation of this unceasing activity. The most powerful computers are pale imitations of the complexity of living organisms, but the fact that computers work at all, as reliably as they do, day after day, makes DNA replication easier to believe. It is perhaps no surprise that our understanding of molecular biology has advanced hand in hand with computer technology; the mind needs familiar metaphors to make its way in the unfamiliar world.

A recent issue of the journal *Science* was devoted to progress in understanding DNA replication.[3] The articles are technical, but it is clear even to a casual reader that molecular biologists are making extraordinary progress toward unraveling this central feature of the riddle of life. To paraphrase Winston Churchill: We are not at the end of understanding, we are not even at the beginning of the end, but we are at the end of the beginning. I was struck by how often the authors of the articles used mechanical metaphors. "Cellular machinery." "Molecular machinery." "Molecular machines." "Molecular motors." "Replication mechanisms." "Mechanisms for maintenance of DNA ends." And so on.

The machine metaphor has been deeply ingrained in scientific thought since the seventeenth century—in astronomy, terrestrial physics, and biology—and retains its hold on our imaginations. But it is not a clockwork of cogs and levers that best describes the machinery of life; rather, it is the buzzing information hive of the silicon chip. An electronic computer is closer in size and speed to what happens inside a living cell; as we pack more and more transistors onto chips, the scale of computers more closely approaches the nanometer scale of DNA. Computers have become indispensable to molecular biologists, not only for building mathematical representations of what happens in a cell, but also as powerfully evocative metaphors.

However, many of us instinctively recoil from the mechanical metaphor for life, and especially for consciousness. We are put off by the idea that we might be merely machines. We cling to the notion that there is something magical, irreducible, and transcendent about life, something that will forever escape the molecular biologists with their computer models of chemical structures. Two things to keep in mind: (1) "Life is a machine" is only a metaphor. As we move away from the plane of immediate sensation, all knowing is metaphorical—in science, as in poetry and art. The mechanical metaphor replaced the animistic metaphor (the attribution of living souls to plants, inanimate objects, and natural phenomena) in science because it usefully extended our mastery of nonhuman nature, mainly in the way it lent itself to mathematical representation. (2) The mechanical metaphor for life or consciousness does not diminish our estimation of our selves. We are what we are what we are.

Mental maps and metaphors come and go; we are still faced with what William Faulkner called "the old verities and truths of the heart, the old universal truths lacking which any story is ephemeral and doomed—love and honor and pity and pride and compassion and suffering."[4] Our science is but a patch on the infinitude of our ignorance. We cannot live without knowledge, but neither is our knowledge exhaustive. The indispensably useful mechanical metaphor for life does not so much reduce the miraculous to the mundane as it elevates the mundane to the miraculous.

Quivering Citadels There

No mental map of the world can exhaust the richness of reality, other than a map that is identical to the world itself, like the map of Swift's story on a scale of one-to-one that must be kept folded lest it completely cover the landscape it describes. Mechanical metaphors, computers, mathematics, and quantitative experiments help us move away from the world of immediate perceptions and reliably explore the universe of the galaxies and the DNA. We may be frightened or depressed by the progress science has made in understanding life, and especially by the effectiveness of the mechanical metaphor, but few of us deny that the advance of molecular biology has been spectacular; it is the crowning intellectual achievement of our century. The mechanical metaphor remains, for better or worse, thc most fruitful way of doing biology.

But we are not merely machines. We are not *merely* anything. The universe is possibly infinite. Our maps of the world are finite, tentative, evolving. There may be new and better descriptions of life lurking in our future, new and more fruitful metaphors. Huge tracts of the biological landscape remain to be adequately mapped—for example, the genesis of the organism from a single cell, and human consciousness—but nothing we have learned about life yet recommends jettisoning the mechanical metaphor. And nothing we know about life requires the existence of a disembodied vital force or immaterial spirits, or a special creation of species.

Are we diminished by this new knowledge? One might as well ask if we are diminished by the expulsion of the fairies from their hills, or of the dryads and naiads from their trees and brooks. We are the same *selves* we always were, with the same loves, hates, dreams, and terrors as when we thought ourselves to be the playmates of gods, immersed in spells and magic. Scientific description does not diminish reality. In the last analysis, we are, as Kierkegaard said, "alone before the face of God, alone in this tremendous exertion and this tremendous accountability."

Our estimation of our self-worth is lodged within ourselves. Nevertheless, it would be disingenuous to suggest that our new scientific maps of the world do not have consequences in our lives. Most especially, the exclusion of the construct "immaterial spirits" from our maps of the world adds the possibility of death's grim finality to the list of problems with which we must learn to cope. On the other hand, the fruitfulness of the mechanical metaphor for life has broadened our understanding of life and made us less fearful and superstitious. Because we no longer believe in fairies, we are free to walk the high road at night, unafraid under a canopy of stars, singing with Hopkins:

Look at the stars! Look, look up at the skies!
O look at all the fire-folk sitting in the air!
The bright boroughs, the quivering citadels there!
The dim woods quick with diamond wells; the elf-eyes![5]

Deep and inviting, beautiful and mysterious, the starry night draws us into communion with a soul and life force greater than ourselves that animates the spiraling galaxies and untangles the knots of DNA.

THREE

The Known and the Unknowable

Penetrating so many secrets, we cease to believe in the unknowable. But there it sits nevertheless, calmly licking its chops.

H. L. Mencken

A *Boston Globe* reader sent me a letter, saying: "I suspect that science makes scientists happier than it does other people. Scientists always realize how much remains to be explained, they realize the limits of scientific explanation, and they have the fun of doing science. To laymen, science often seems to take away mystery and make them feel a little stupid at the same time. . . . It seems to me that there is a growing suspicion that while science might be useful it is also spiritually destructive. A lot of people clearly want to feel that there are things out there that can't be explained. Maybe we all feel that way to some extent."

Yes, I suppose we all do feel that way to some extent. The human mind loves a mystery, loves a world possessed by spirits. Alas, scientific skepticism has a way of debunking the spirits. The gods have been tossed from their Olympian thrones, the spirits of trees and brooks sent packing. Is there nothing, then, that Skeptics will not profane? Like a heartless landlord who cares more about profit than compassion, science has evicted the fairies from their hills, hell-bent upon rooting out mystery.

But before we abandon Skepticism for True Belief, and turn to horoscopes, parapsychology, New Age superstitions, and religious fundamentalism, let me speak for science. Science is boring by design. Scientific communication has evolved a style that is deliberately devoid of passion, poetry, and the longings and despairs of the human heart. Why? To get on with the business of finding out how the world works. Science is the one human endeavor that has proven relatively immune to the passions that divide us. There is no such thing as Jewish science, Christian science, Muslim science, Buddhist science. There is no such thing as male or female science, black or white science, Democratic or Republican science. This is not to say that individual scientists can't be sexist, racist, or politically committed, or that science itself hasn't been shaped by its Judeo-Christian, European, male-dominated origins. But by keeping, as best we can, human differences out of the communication of science, we have forged a tool for human improvement that is anchored in repeatable, verifiable observation, rather than in prejudice and passionate conviction.

And, yes, the human improvement is there. We no longer lash ourselves in retribution when we see a comet in the sky, or bolt our doors at night against fairies. All things considered, we live longer, healthier, more peaceable and prosperous lives because of science. Of course, we also have atomic bombs, ozone holes, acid rain, and the monstrous tinkerings of genetic engineers, all of which can arguably be laid at the doorstep of science. But how many would prefer to turn the clock back to the Middle Ages? The last witches were burned during Newton's lifetime. The last visitations of the plague in Europe coincided with the Scientific Revolution. A certain element of disenchantment may be the price we pay for freedom from the darker excesses of True Belief.

An Island in a Sea of Mystery

Is science spiritually destructive, as my correspondent suggests? Does it throw cold water on every flickering spark of mystery? In chasing the fairies from their hills, does it leave in its wake a landscape devoid of spirit? Is science the enemy of soul? In an earlier book, *Honey from Stone*, I proposed the following metaphor:

> Let this, then, be the ground of my faith: All that we know, now and forever, all scientific knowledge that we have of this world, or will ever have, is as an island in the sea [of mystery]. . . . We live in our partial knowledge as the Dutch live on polders claimed from the sea. We dike and fill. We dredge up soil from the bed of mystery and build ourselves room to grow. And still the mystery surrounds us. It laps at our shores. It permeates the land. Scratch the surface of knowledge and mystery bubbles up like a spring. And occasionally, at certain disquieting moments in history (Aristarchus, Galileo, Planck, Einstein), a tempest of mystery comes rolling in from the sea and overwhelms our efforts, reclaims knowledge that has been built up by years of patient work, and forces us to retreat to the surest, most secure core of what we know, where we huddle in fear and trembling until the storm subsides, and then we start building again, throwing up dikes, pumping, filling, extending the perimeter of our knowledge and our security.[1]

Knowledge is an island in a sea of mystery. The metaphor takes its power from a firmly held fact: We live in a universe that is infinite, or effectively so. Our brains are finite, a mere 100 billion nerve cells. Our mental maps of the world are therefore necessarily finite. As time passes, the scale and detail of our maps increase, but they no more exhaust the worldscape they describe than a map of the Grand Canyon depletes the power of that natural chasm to astonish and surprise.

If we accept that knowledge is a finite island in a sea of inexhaustible mystery, then two corollaries follow: (1) The growth of the island does not diminish the sea's infinitude, and (2) the growth of the island increases the length of the shore along which we encounter mystery. It is this last aspect of the metaphor that is most important. We are at our human best as creatures of the shore, with one foot on the hard ground of fact and one foot in the sea of mystery. Bureaucrats, technocrats, and scientific drudges keep to the high ground, their noses fixed in ledgers and laboratory notebooks. New Age dreamers flounder in water over their heads, with near horizons. It is at the shore that the creative work of the mind is done—the work of the artist, poet, philosopher, and scientist.

Art and science flourish at the boundary of the known and the unknown. The biologist Erwin Chargaff, who contributed mightily to our knowledge of DNA, wrote: "It is the sense of mystery that, in my opinion, drives the true scientist; the same blind force, blindly seeing, deafly hearing, unconsciously remembering, that drives the larva into the butterfly. If [the scientist] has not experienced, at least a few times in his life, this cold shudder down his spine, this confrontation with an immense invisible face whose breath moves him to tears, he is not a scientist."[2] The writer Vladimir Nabokov advises students in his lectures on literature to trust the spine and its tingle as the most reliable guide to art and science. The shoreline between knowledge and mystery is the place to feel the tingle, where mind and heart are equally engaged by the world's peculiar geography.

A Sea of Starlight

Last night I walked the Kerry shore on the first clear dark night of summer. Arcturus blazed at the zenith among a court of stars. I thought of Shakespeare's "night's candles"; Hopkins's "fire-folk sitting in the air"; Van Gogh's *Starry Night on the Rhone*, those fiery vortices, those furious whirlpools of light. I considered that Haydn's majestic *Creation* oratorio—"the space immense of th' azure sky, a countless host of radiant orbs adorn"—was probably inspired by a view of the stars through astronomer William Herschel's telescope in 1792. These star-struck tingles in the spine are universal, timeless, burning, fierce.

Shakespeare, Hopkins, Van Gogh, and Haydn gave brilliant expression to the power of the stars to elevate and surprise.

For a moment, let me turn the calculating eye of science on those glittering points of light.

Arcturus is thirty-six light-years away. That's 216 trillion miles, yet I somehow manage to see the star! We have seen so many pictures of Stars of Bethlehem and Twinkle Twinkle Little Stars with beams of light shooting straight down to Earth, that it's easy to imagine that the light from Arcturus is somehow directed toward us alone. But the light from a star radiates outward in every direction, stretched thinner and getting weaker as it moves away from the star. Only the tiniest fraction of Arcturus's light falls upon the Earth. How much? Let's do a calculation.

At a distance of 216 trillion miles, the light of Arcturus is spread out over a sphere with an area of 586,000,000,000,000,000,000,000,000,000 square miles. The Earth has a cross-sectional area of about 50 million square miles. So the fraction of Arcturus's light that falls upon the Earth is about one part out of 10 sextillion. (That's the number one followed by twenty-two zeros.) Of the starlight that falls upon Earth, an even tinier fraction enters the pupil of my eye to form an image of the star. How does the area of my pupil compare to the cross-sectional area of the Earth? Another factor of 10,000,000,000,000,000,000, more or less. So the fraction of Arcturus's light that enters my eye is one part out of 100,000,000,000,000,000,000,000,000,000,000,000,000,000.

Let's see if I can give you a sense of what that last number means. The oceans of the Earth contain about 320 million cubic miles of water. If you dipped the tip of a pencil point into the ocean, the amount of water you'd come up with, compared to all the waters of the oceans, is *more* than the fraction of Arcturus's light that enters my eye.

Think about it. All those expanding spheres of light from thousands of stars in the spangled sky (the ones that are bright enough to see), overlapping about through space, falling upon the pupils of our eyes from only slightly different directions—and out of a damp pencil tip's worth of energy, our eyes and brain form an image of Arcturus. And Vega. And Deneb. And Altair. And . . . And that's just the brighter stars. If we consider the most distant stars that we can see with the unaided eye, which would be roughly 10,000 light-years away, then we get another factor of . . .

Stop! I have indulged your patience long enough. But consider how we have lengthened the shore of the island of knowledge and explored new horizons of mystery. We have tangled briefly with a universe that is not merely a projection of ourselves—"fire-folk sitting in the air" or "night's candles"—but something deeper and more complex and infinitely mysterious. We have glimpsed a universe in which stars do not send their beams for our eyes alone. Starlight inflates across light-years to be merely sampled by our eyes—damp pencil tips of energy, but enough to excite the retina, signal the brain, form images, open our minds to the universe of the galaxies, inspire poets and artists, frighten, elevate, surprise, and ignite the shudder in the spine.

Something Subtle

Are science and the aesthetic emotion at odds? I am reminded of something the late great physicist and irrepressible raconteur Richard Feynman said in an interview with his biographer, Christopher Sykes. He described an artist friend who would hold up a flower and say: "I, as an artist, can see how beautiful a flower is. But you, as a scientist, take it all apart and it becomes dull."[3] Of course, Feynman's friend is evoking the age-old bugbear of the romantic, Emily Dickinson's "monster with a glass," William Wordsworth's "meddling intellect" who "murders to dissect." Feynman responds:

> First of all, the beauty [the artist] sees is available to other people—and to me too. Although I might not be quite as refined aesthetically as he is, I can appreciate the beauty of a flower.

> At the same time, I see much more about the flower than he sees. I could imagine the cells in there, the complicated actions inside, which also have a beauty. I mean, it's not just beauty at this dimension of one centimeter: there is also beauty at a smaller dimension—the inner structure. The fact that the colors in the flower are evolved in order to attract insects to pollinate it is interesting—it means that the insects can see the color. It adds a question: does this aesthetic sense also exist in the lower forms? Why is it aesthetic? All kinds of interesting questions which a science knowledge only adds to the excitement and mystery and the awe of a flower. It only adds. I don't understand how it subtracts.

Feynman is playing on the shore of knowledge and, in classic Feynman fashion, having fun. He doesn't exactly know what he is talking about (he is a physicist, not a molecular biologist or chemist), but he knows enough to tease out a profound secret regarding the unity of life.

> Does it make any less of a beautiful smell of violets to know that it's molecules? To find out, for example, that the smell of violets is very similar to the chemical that's used by a certain butterfly . . . to attract all its mates? It turns out that this chemical is exactly the smell of violets with a small change of a few molecules. The different kinds of smells and the different kinds of chemicals, the great variety of

> chemicals and colors and dyes and so on in plants and everywhere else, are all very closely related, with very small changes, and the efficiency of life is not always to make a new thing, but to modify only slightly what is already there, and make its function entirely different, so that the smell of violets is related to the smell of earth.

Feynman is not dismissing art, nor is he suggesting that science is somehow more important than art. He is merely reacting in his characteristically uninhibited way to a cranky opinion of his artist friend. Art and science are each sublime activities of the human mind; we are less than human without either. Both activities thrive at the interface of knowledge and mystery. Each is enriched by the other, and the two activities, although different, have more in common than either scientists or artists are usually willing to admit.

In some ways, science has a more modest goal than art. It moves forward by measured degrees, nibbling at mystery. It does not aspire to Truth, only tentative, incremental truth. It sacrifices the capital *T* for confidence, reliability, progress, power. Science dreams of objectivity, even if it means focusing on that limited part of human experience that is amenable to logical analysis. It is willing to temporarily suppress part of what it means to be human in order to gather to itself more reliable knowledge of the nonhuman world.

Art's purpose is bolder. Art trusts intuition to grasp great Truths, even if only as chimeras and ghosts. It is willing to sacrifice quantitative certainty, public consensus, and practical power for spiritual union with the thing experienced. It bends the world to human purpose. It does not pretend to be objective.

The scientist who does not allow herself to be spiritually empowered by art is the poorer for it. And the artist who dismisses science has closed himself off from half of the human adventure. As Feynman says, with his usual mischievous grin, "It's much more wonderful to know what something's really like than to sit there and just simply, in ignorance, say, 'Oooh, isn't it wonderful!' "

Scientists are not often adept at communicating the human import of their endeavors. The deliberately dispassionate and tediously quantitative language of science is not easy to master; once mastered, it is not easy to put aside. We need more scientists with Feynman's wit and common touch to plug us into the universe of the DNA and the galaxies, and more poets who can absorb the knowledge of science and infuse it with human passions.

At its best, scientific skepticism is a manifestation of curiosity, intelligence, and imagination—in a word, the best of the human *spirit*. It slowly, patiently builds the domain of knowledge, pushes back the encroaching darkness, the demons of the deep, but never exhausts the infinitude of mystery. Asked if he was religious, Einstein replied: “Try and penetrate with our limited means the secrets of nature and you will find that, behind all the discernible concatenations, there remains something subtle, intangible and inexplicable. Veneration for this force beyond anything that we can comprehend is my religion.” Call it, if you want, God.

FOUR

Organized Skepticism

Science frees us in many ways . . . from the bodily terror which the savage feels. But she replaces that, in the minds of many, by a moral terror which is far more overwhelming.

Charles Kingsley (1866)

IN THE AUTUMN OF 1995, the Hubble Space Telescope research team released a spectacular photograph of a star-forming region in the constellation Serpens. I saw the photo on the evening news and immediately downloaded it over the Internet onto the screen of my computer. Breathtaking! A luminous cloud, called the Eagle Nebula, 7,000 light-years away and tens of trillions of miles wide. Three tall columns of glowing gas, a cosmic coral reef, emerald tinted, incandescent. At the top of the tallest column, rays of light stream outward from the hot interior, blowing away outer layers of the cloud, except where newly formed stars, hidden in their swaddling wraps, hold the gas in place. An evaporating stellar nursery, revealing infants in the nest—new suns, planet systems, worlds. We had seen the Eagle Nebula before, particularly in the magnificent photograph made by Ray Sharples with the Anglo-Australian Telescope at Siding Spring, Australia. But the Hubble picture showed fresh details, as if we were seeing the leaves on a tree for the first time. Looking at the new image—the streaming light, the dark globules of condensing stars—we felt like witnesses to the creation.

Sitting awestruck before the computer screen, I did not forget that the photograph was made by an instrument of human invention, orbiting the Earth in space and directed with precision by astronomers on the ground. I was viewing the photograph on a high-resolution color monitor connected to the Space Telescope Science Institute by a lightning-fast stream of electronic bits. With the click of a mouse, I was able to plug my imagination into a stellar nursery many thousands of light-years from home.

Yearning and Learning

A few days later, I read in the newspaper that when the photograph was shown on CNN, the network was flooded with calls from viewers claiming to have seen the face of Jesus in the billowing cloud. Here is a classic manifestation of True Belief: seeing what one wants to see, reaching for human acknowledgment in a distant nebula. I pulled up the image again onto my computer screen and looked carefully, squinting my eyes, turning my head, sideways, upside down. I saw what appeared to be the face of a gorilla—King Kong, perhaps—in the tallest column of the nebula. As for Jesus, I couldn't find him anywhere.

Why do so many of us see the face of Jesus in a nebula, sea monsters in Loch Ness, flying saucers in the sky? Why do we see drops of blood streaming from the painted wounds of a crucifix, the spinning Sun at Fátima, canals on Mars? We *yearn* to be part of something greater than ourselves. We *learn* by hard experience that miracles don't happen. Yearning and learning are integral parts of being human (the first may be genetic; the second we must work at). We cannot be fully human without both. Finding the proper balance between yearning and learning can keep us occupied for a lifetime.

We yearn when we dream of fulfillment, of greater happiness, of knowing more. We yearn when we cry out for human affirmation from the cosmos, when we love, when we laugh, when we cry, when we pray. Yearning is wondering what is around the next bend, over the rainbow, beyond the horizon. Yearning is curiosity. Yearning is the driving force of science, art, and religion.

Learning is listening to parents, wise men and women, shamans. Learning is reading, going to school, traveling, doing experiments. Learning is dismantling the clock to see what makes it tick or touching the stove to see if it's hot, not taking anyone's word for it (not even the word of parents, wise men and women, shamans). In science, learning means trying as hard to prove that something is false as to prove it true, even if that something is a cherished belief.

Yearning without learning is seeing Elvis in a crowd, the fossilized footprints of humans and dinosaurs together in ancient rocks, or moving statues. Yearning without learning is buying tabloid newspapers with headlines announcing "Newborn Baby Talks of Heaven" and "Aliens in U.S. Congress!" Yearning without learning is looking for healing in pretty crystals and the meaning of life in horoscopes. Yearning without learning is following whatever current guru offers the most promising prospects of eternal life.

Learning without yearning is pedantry, scientism, idées fixes. Learning without yearning is believing that we know it all, that what we see is what we get, that nothing exists except what can be presently weighed and measured. Learning without yearning is rote science without a heart, without a dream, without a hope of beauty.

Yearning without learning is seeing the face of Jesus in a gassy nebula. Learning without yearning is seeing only the gas.

What Is an Angel?

In his book *Leaps of Faith,* British psychologist Nicholas Humphrey has written brilliantly on the motivations of the True Believer. All supernatural belief systems have catered to two central needs, he says: the need for rational understanding of the surrounding world, and the need for emotional security within it. These correspond, more or less, to what I have called learning and yearning. Humphrey draws our attention to the three questions scrawled in the corner of Paul Gauguin's last painting: *D'où venons nous? Que sommes nous? Où allons nous?* From where did we come? What are we? Where are we going? These are the same questions that were asked on the first page of my childhood catechism. Who am I? Where did I come from? Why am I here? People want *explanations* for the first two questions, says Humphrey. They want *reassurances* for the third.[1]

The answers I learned as a child were part of a neat package. God made me. At the moment of conception, he breathed life into a tiny lump of inert matter, thereby creating an immortal soul, my self, unique from every other being. He made me to know him, to love him, and to serve him. And my fate? Everlasting bliss in his presence. If I was good, that is. Otherwise I could look forward to eternal torment in the fires of hell.

Part of being good was buying the package whole, the entire catechism. One was not allowed to pick and choose. Doubts were not admitted. Call any part of the system into question, and the whole thing was in danger of coming apart, because, to tell the truth, none of it was based on the kind of evidence that might impress a scientist, a court of law, or even a reasonably skeptical child. The legitimacy of the system was guaranteed by revelation and holy tradition, the sources of which were conveniently tucked away in the past, beyond immediate inspection. For assurances of the veracity of the sources, we had to rely on the magisterial authority of the Church, as represented by priests, nuns, bishops, and pope. That authority was divinely inspired, infallible.

Religion, therefore, was a bit of a pig in a poke. It was a comprehensive system, with tentacles reaching from the great octopus-head of revelation into theology, philosophy, art, architecture, music, literature, history—even into the rhymes and games of youth. Powerful, compelling. The whole thing stood or fell on a single premise: eternal salvation. Believe, and you shall be saved. Buy the pig in the poke, and death will have no sting. Anything that appeared to confirm the supernatural order made the system more tenable. Guardian angels, devils, stigmatas, apparitions, levitations, miraculous cures—all buttressed the system. Even secular manifestations of the paranormal (mind readers, poltergeists, the evil eye) were secondhand evidence for the promise. The material world, the world accessible to scientific study, was believed to be a dreary shadow of the *real* world of spirits, powers, invisible presences, immortal souls.

Supernatural belief has much to recommend it. Religion in particular provides a sense of belonging to a group, a history, and a culture in which to take pride, great works of art, stirring literature, service to the poor and needy, satisfying liturgical celebrations of creation, and rites of passage. Looking back, I hanker for the cozy securities, the certainties, the cocooning warmth of the candlelight, incense, and Gregorian chant, the consolation of belonging to the one true faith, of being an *insider*.

We were told nothing in our religious education about the anthropological foundations of religion, comparative religion, or the terrible atrocities and genocides that had been perpetrated in the name of our triumphalist faith. We heard nothing about the many contingencies of Church history, the abominable behaviors of Renaissance popes, the egregious materialism of the Vatican. It was never suggested that one might choose to live ethically without the threat of hellfire, or that the tendency toward altruism might be part of the genetic inheritance of every human being.

In our science classes in Church-sponsored primary and secondary schools, we learned innocuous facts, the bare rudiments of biology, chemistry, and physics, but never the grand syntheses—celestial mechanics, geological chronology, natural selection, molecular biology—never the vibrant, resonant web. We were never given a hint of what science really was: organized skepticism. The postmedieval cosmic order was kept firmly out of sight.

Later, as a university undergraduate, I fell into a dreamy delirium of neopagan Catholicism. I reveled in the splendidly embroidered rhythms of the canonical hours and liturgical year. I read those twentieth-century Catholic (or ex-Catholic) authors who pitted the soul against Satan in a life-or-death struggle: Bernanos, Bloy, Mauriac, Péguy, Kazantzakis, Greene. In fits of asceticism, I put pebbles in my shoes and sand in my bed. Under the influence of Thomas Merton's *Seven Storey Mountain*, I made my way to the gate of the Trappist monastery in Gethsemane, Kentucky, arriving, appropriately, in the middle of the night. I was yearning, with all my heart, to be raptured, transported, lifted up and out of myself—by God or the devil, it didn't matter.

Meanwhile, I was learning science, real science—discovering the power of empirical thinking and catching a glimpse of the God of creation. I determined to examine afresh the dogmas of my faith. I decided I would read the recently published *Hawthorn Twentieth Century Encyclopedia of Catholicism.* This many-volumed series of small books, each about one aspect of the faith, was purportedly scholarly and up-to-date. The first volume I picked up was *What Is an Angel?* A few dozen pages into the book, it dawned on me that if I could believe in angels, I could believe in anything. And I didn't believe in angels. The evidence for angels was as convincing as the evidence for poltergeists, fairies, or Bigfoot; perhaps less convincing. I closed the book, and the whole cosmological system of my childhood came tumbling down. I had become a Skeptic.

Confirming the Dark Immensities

In *Feet of Clay,* British psychologist Anthony Storr writes about the appeal of supernatural belief systems as proffered by charismatic gurus from Jesus to the Bhagwan Shree Rajneesh. He believes it is impossible to understand the power of gurus over our emotional and intellectual lives unless we rethink conventional notions of madness and sanity. "Idiosyncratic belief systems which are shared by only a few adherents are likely to be regarded as delusional," he writes. "Belief systems which may be just as irrational but which are shared by millions are called world religions."[2] I think of some of the irrational stuff I was asked to believe as part of the cosmological foundations of my Roman Catholic faith—angels, exorcisms, heaven, hell, purgatory, and indulgences—and I'm inclined to believe that Storr is right.

American culture at the end of the millennium is in the thrall of entrepreneurial gurus—New Age and fundamentalist—who promise bliss. They know our weakness. We hanker for the unseen world of powers and spirits because we are frightened by the apparent ephemerality of the material world. Three hundred years ago, Pascal eloquently described this fear: "When I consider the brief span of my life, absorbed into the eternity before and after, the small space I occupy and which I see swallowed up in the infinite immensity of spaces of which I know nothing and which know nothing of me, I take fright. . . . The eternal silence of these infinite spaces terrifies me."[3] Science confirms the dark immensities, the silent infinitudes. No wonder so many people reject scientific cosmology. No wonder so many people saw the face of Jesus in the nebula. It is our own face we wish to see there—abiding, undissolvable, cosmic.

Everything we have learned in science since the time of Galileo suggests that the nebulas and galaxies are oblivious to our fates. Everything we have learned suggests that our souls and bodies are inseparable. Everything we have learned suggests that the grave is our destiny. Therefore, if the promise of eternal life is to have maximum drawing power, it is essential for Church and guru to undermine the legitimacy of science.

A Closed Establishment?

My newspaper column often evokes responses from readers who seek to convince me that science is flawed at its core. They are troubled by inconsistencies between scientific cosmology and religious cosmology. Their protests usually purport to show that evolution is wrong, that the Earth is younger than we think, that the Big Bang never happened (how could everything have come from that tiny point?), or that the evidence for supernatural phenomena is irresistible. These offerings range from the terribly clever to the merely silly. They invariably protest against the "close-mindedness" of the scientific establishment. If only scientists would open their eyes, these True Believers say, the evidence for the supernatural would be staring them in the face.

Are my complainants onto something? Is science implacably opposed to religion? Can evidence for the supernatural receive a fair hearing in science? Or is science locked in an ironbound orthodoxy (atheistic, secular humanistic, Darwinian, materialistic) that admits no breach of faith?

Science *is* exclusionary. If every idea has equal currency in the marketplace of ideas, then truth becomes a matter of whim, politics, expediency, or the tyranny of the strong. Science has evolved an elaborate system of social organization, communication, and peer review to ensure a high degree of conformity within an institutionally supported orthodoxy. This conservative approach to change allows for an orderly and exhaustive examination of fruitful ideas. It provides a measure of insulation from fads, political upheavals, religious conflicts, and international strife. Yes, offbeat ideas do have a hard time of it in science. But not an impossible time. Revolutions in science are few and far between, but they do happen. Science is conservative, but of all truth systems that purport to explain the world, it is also the most progressive.

Alan Lightman of the Massachusetts Institute of Technology and Owen Gingerich of the Harvard-Smithsonian Center for Astrophysics, writing in the journal *Science*, point out that scientists may be reluctant to face change for the purely psychological reason that the familiar is more comfortable than the unfamiliar.[4] But Lightman and Gingerich also recognize that a conservative system of truth provides an efficient framework for assimilating the multitude of facts that scientists observe. They are particularly interested in the fate of "anomalies," observations that don't fit the accepted scientific orthodoxy. From the history of astronomy, geology, and biology, they offer examples of exceptions to a prevailing theory that were simply ignored by scientists.

For instance, for centuries before Darwin, natural philosophers marveled at the exquisite adaptation of organisms to their environment, which they took as evidence for intelligent design. Camels carry their energy-storing fat in one place, on their backs, so that the rest of their bodies can efficiently cool off in the deserts where camels live. Giraffes have long necks that allow them to eat from the high ungrazed trees of the savanna. And so forth. This specificity of design was thought to be compelling evidence for the work of an intelligent Creator, as described in Genesis; indeed, a wonderful degree of *suitableness* does exist almost everywhere in nature.

But what about birds, such as the ostrich, that have wings but do not fly? Why do blind fish that live in lightless caves have eyes? What might an Intelligent Designer have had in mind? These examples of apparently maladaptive design were ignored by scientists until Darwin proposed a new theory—natural selection and common descent—that explains with equal facility the hump of the camel, the neck of the giraffe, the wings of the ostrich (descended from birds that flew), and the eyes of the blind fish (descended from fish that lived in light). Anomalies are usually acknowledged only in retrospect, say Lightman and Gingerich. When a new and more inclusive theory gives an explanation of previously unexplained facts, it becomes "safe" to recognize anomalies for what they are. In the meantime, scientists ignore what doesn't fit.

"Exactly!" cry creationists, supernaturalists, and paranormalists. "Scientists ignore what doesn't fit." "Scientists work with blinders on their eyes." "Science is an orthodoxy more rigid than the most conformist religion." Well, yes and no. Science *is* conservative, as it must be if it is to provide a stable framework for understanding the world. But I know of no scientist who does not admit that our present understanding of the universe is tentative and incomplete. Even cherished ideas have been overthrown when the pressure for change becomes irresistible—witness recent revolutions in geology and cosmology that cast aside firmly held beliefs in fixed continents and a steady-state universe. Creationists and advocates of the paranormal emphasize the anomalies in science and ignore the vast system of mutually supporting ideas. Scientists focus on the broad orthodoxy and temporarily ignore the intractable exceptions. Perhaps neither attitude toward observations is perfect, but the latter is certainly the more fruitful.

Like any human activity, science is subject to the

prejudices and personality defects of individuals; scientists can indeed be arrogant and close-minded. But the scientific community has evolved a system of checks and balances that ensures a fair hearing for any potentially fruitful idea, at the same time holding at a distance the tide of bogus opinion that constantly laps at the shores of legitimate public knowledge. The system occasionally slips, but science is the one truth system committed to change rather than preservation. Ironically, some measure of conservatism may be the best way to ensure that progress is made.

Walking on Fire

The nineteenth-century physicist Michael Faraday once said, "Nothing is too wonderful to be true." With that in mind, the Skeptic must be open to the possibility that an apparently offbeat idea contains a germ of truth. At the same time, he is right to insist that certain evidential criteria must be met for an idea to qualify as science. If science is open to every private vision of reality, then its usefulness as public knowledge is severely impaired.

But what about irrational assertions that are embraced by millions and are the tenets of the world's great religions? According to a 1995 Gallup poll, nearly three-quarters of Americans believe in angels, despite the fact that not a shred of scientific, reproducible, nonanecdotal evidence can be adduced in their support. Believers in angels will suggest that it is the nature of angelic intervention in the affairs of humans *not* to be reproducible, or that angels choose to manifest themselves only to Believers. And what can the Skeptic counter to *that*? Either one believes, or one doesn't. The Skeptic looks at the big picture (the fabulously successful web of scientific knowledge) and ignores the occasional exceptions (anecdotal accounts of angels). He does not a priori rule out the existence of angels, but he knows of no evidence for their existence that cannot be more economically explained by more mundane hypotheses. The Skeptic's guiding principle is Ockham's razor: No more things should be presumed to exist than are necessary to explain the phenomenon.

Consider the newest fad in our search for pop spirituality: fire walking. A new breed of entrepreneurial guru stands ready to lead us to a life beyond merely material existence—across twelve feet of glowing coals. The typical fire walk goes something like this. A few dozen people pay fifty dollars or more to participate in a fire walk seminar. After an introductory pep talk from the instructor, they are led outside, where half a cord of blazing wood is spilling sparks into the night sky. (Fire walks generally take place in the evening so that the light of the fire and, later, the glowing coals will be more spectacular.) Then, back inside for a couple of hours of consciousness-raising mumbo jumbo: "the power of mind over matter," "life energy," "body auras," "fields of consciousness"—that sort of thing. The laws of physics are made to be broken, says the instructor, if only we can harness the spiritual power that lies deep within us.

By now the fire has burned down to a pile of glowing ash, with a temperature of more than 1,200 degrees Fahrenheit. The participants are led outside again, barefoot, chanting confidently, in a state of high excitement. They hold hands in a circle as the coals are raked into a long, narrow path. The leader takes a deep breath and strides across the coals. One by one the participants follow, down the glowing path, into the congratulatory arms of their fellow fire walkers. They have participated in a "miracle."

Now, don't get me wrong. I am not putting down the fire-walking experience itself. Fire walking has a long history within certain cultural traditions as a religious ritual or rite of passage; in this regard, fire walking can convey the same sacramental symbolism as the bread of the Eucharist or the waters of baptism. As an extreme sport, fire walking ranks right up there with sky diving and bungee jumping. As a sort of quickie self-help program—overcoming fear, developing self-confidence—it might also have something to recommend it. But a "miracle"? Hardly. Mind over matter? Not a chance. A violation of the laws of physics? Never. In fact, it is *because* of the laws of physics that fire walking is possible.

The heat capacity of wood ash is small. Although the temperature of the glowing coals is high, the amount of heat contained within them is deceptively low. (The same is true for the air in a hot kitchen oven, which is why you can safely put your hand in the oven.) Also, the thermal conductivity of wood ash is low. During the fraction of a second that the foot is in contact with the coals, there is not enough time for a damaging amount of heat to transfer to the skin.

Fire-walking gurus dismiss this physical explanation as typical close-minded skepticism. Scientists will try to explain away any phenomenon that doesn't fit the materialist dogma, they say. One fire-walking enthusiast writes, "The more fully we have adopted a posture of skepticism, the more difficult it becomes to approach anything in life with the attitude and posture of open faith and trust."[5] And he is partly right. Skepticism *by itself* is sterile. Skepticism *by itself* can be arrogant. Skepticism *by itself* can close the door to new experience. But scientific skepticism is coupled with another principle: Ockham's razor. If something can be explained simply, in a familiar way, then it is best to avoid more exotic explanations. If the thermal properties of wood ash explain fire walking, then there's no need to invoke auras and spirits.

In the fall of 1996, I wanted to write about fire walking in my *Boston Globe* column. I wanted to write about the thermal properties of wood ash, Ockham's razor, and the superfluousness of "mind over matter" and "fields of consciousness" for explaining why it is possible to walk unscathed over hot coals. But I knew I could not do so credibly without putting my own feet to the fire. Of course, one should not try fire walking without expert guidance or knowing exactly what one is doing; it is possible to get seriously burned. Fire-walking gurus require their customers to sign waivers of responsibility. Nevertheless, one Sunday afternoon I built a bonfire in my backyard and raked the glowing coals into a two-foot square. While witnesses watched, I stepped barefoot onto the red-hot coals, then again, and again, and again. No burns. No blisters. Can't even say that I felt anything unusual. But I will admit that the first step was scary. My successful fire walk was not mind over matter, but mind over mind. A small victory for Ockham's razor.

Of course, True Believers will say that my feet were protected from burns by my own involuntary powers of mind. No talk of thermal physics will dissuade them from their belief. Over 100 years ago, Roman Catholic True Believer John Henry Newman offered this profession of his faith: "[The Believer] is sure, and nothing shall make him doubt, that if anything seems to be proved by astronomer, or geologist, or chronologist, or antiquarian, or ethnologist, in contradiction to the dogmas of faith, that point will eventually turn out, first, *not* to be proved, or secondly, not *contradictory*, or thirdly, not contradictory to any thing *really revealed*, but to something which has been confused with revelation."[6] It is a wonderfully flexible formula, allowing the True Believer to have his ideological cake and eat it too.

There was a time when the Christian Church insisted that heliocentricity was inconsistent with revealed truth. Eventually the evidence for the motion of the Earth about the Sun became so overwhelming that the guardians of dogma were forced to declare that what had previously been thought to be literally true (the scriptural account of Joshua making the Sun stand still) was merely figurative; what was *thought* to be revelation wasn't revelation at all. Similarly, it was once preached from pulpits that God sent the Black Death to scourge a faithless people; when it was discovered that fleas on rats were the agents of the disease, it was still possible to claim that God sent the rats. When the rats (and the disease) were abolished by effective public sanitation, then God presumably found other ways of punishing faithlessness. I'm taking cheap shots, of course, but traditional religious faiths, like New Age paranormal fads, have a history of evoking one or the other of Newman's convenient "outs" as scientific truth advances.

It is part of the Skeptic's position that our knowledge of the world is incomplete. This incompleteness will always allow an "out" for the True Believer. Powers of mind over matter can be demonstrated to be *superfluous* for successful fire walking, but they cannot be *disproved.* If scientists don't know everything, then the disembodied soul (for example) is free to roam the vast territories of our ignorance. According to a 1995 Gallup poll, nine out of ten Americans believe in heaven. Three-quarters believe in hell. And why not? The alternative to belief is Pascal's "infinite immensity of spaces of which I know nothing and which know nothing of me." Science cannot rule out heaven or hell because they are beyond the reach of empirical investigation. And if Jesus appears in the swirling gas of a distant nebula, then—as True Believers have suggested—it confirms the hopeless inadequacy of science to capture the suprasufficiency of the world.

FIVE

Astrology and Prayer

What a man would like to be true, he preferentially believes.

Francis Bacon (1620)

In supermarket tabloids, popular magazines, and fundamentalist tracts, we hear anticipations of Armageddon, a crescendo of loony superstition and paranormalism that will culminate in apocalyptic hullabaloo in the last days of the year 1999. Of course, because there is no year zero, 2,000 years of the Christian calendar will actually have passed at midnight, December 31 of the year 2000, not at the end of 1999. But arithmetic is irrelevant. Our frenzied approach to the millennium has nothing to do with mathematics or calendrical science. We are talking about habits of the human mind—about yearning unqualified by learning—and psychologically speaking, the moment of consequence occurs when the triple nines roll over all at once to become born-again zeros.

To be sure, not everyone expects the Day of Wrath. But those who want to believe that the end of the world is at hand find ample evidence for their belief. Floods, hurricanes, volcanic eruptions, comets, and eclipses are taken as signs of the Second Coming. If one is looking for portents, nature will always oblige. Coincidence is the science of the True Believer.

In the spring of 1996, a religious panic swept devout Roman Catholic families in Colombia. Word went around that any soul not baptized by June 6, 1996—the sixth day of the sixth month of a year ending in six (666 is the biblical number of the beast)—would be claimed by the Antichrist on the Last Day. According to news reports, parents thronged to churches with their unbaptized children, clamoring for the sacrament, while Church officials struggled to suppress hysteria. But triple sixes can't hold a candle to the end of a millennium. The cheap little buzz we get from watching the odometer on our automobile hit 10,000 is amplified to cosmic proportions by millennial madness. The arrival of the three big goose eggs on the world's odometer must surely presage the apocalypse—or if not the apocalypse, then at the very least an alien invasion.

Why are so many people irrationally fearful of the millennium? Many of us are willing to surrender rationalism for emotional affirmation, even if the emotional affirmation comes with a frisson of fear. The one thing all superstitions and pseudosciences share is a central significance of the self. We need to feel that the cosmos takes notice of our personal existence, that we are not merely dust motes dancing in the infinite immensity of spaces, in the cavernous galactic silences. "Inspect every piece of pseudoscience and you will find a security blanket, a thumb to suck, a skirt to hold," wrote the science writer Isaac Asimov. He added by way of explanation: "What have we [in science] to offer in exchange? Uncertainty! Insecurity!"

Science and Pseudoscience

Quick. Give brief definitions of *ESP* and *PCR*. What can you say about *Yeti* (Bigfoot) and *SETI*? What are a *UFO* and a *WIMP*?

Likely, you got the first terms of these pairs, but not the second. The first are drawn from pseudoscience; the second, from science. Don't be embarrassed, even if you count yourself a Skeptic; most of us would respond the same way. On a visit to one of Boston's elite high schools, I listed twenty terms on a blackboard—the six above and the following: *cosmic microwave background radiation, Adam and Eve, parallel processing, cellular automata, reincarnation, Loch Ness monster, hot fusion, superconductivity, Shroud of Turin, close encounters of the third kind, the Genome Project, the Bermuda Triangle, scanning tunneling microscope,* and *horoscope.* I went through the list one by one, asking for a show of hands on the following question: "For which topics could you give a reasonably confident explanation?" You can guess the outcome. For half of the terms, most hands in the room were raised. For the other half, only one or two hands went up. I'm sure you can guess which terms fell into each category.

Half of the terms are deemed bogus by the scientific community. The other half are drawn from cutting-edge contemporary science. The latter refer to important developments that have the potential to profoundly change the way we live or the way we understand our world. I have no doubt that the exercise would have produced the same result with students at Harvard or almost any other audience not made up of professional scientists. A 1989 survey showed that about 6 percent of adults in the United States can be called scientifically literate.[1] Skeptics and True Believers alike are vastly less knowledgeable about science than about pseudoscience and superstition.

PCR (polymerase chain reaction) provides a way to replicate tiny quantities of DNA millions of times over; it may be the most important invention of the twentieth century. SETI (search for extraterrestrial intelligence) may answer the question: Are we alone in the cosmos? WIMPs (weakly interacting massive particles) are prime candidates for the so-called dark matter, which might constitute as much as 95 percent of the universe. Cosmic microwave background radiation is the flash of the Big Bang, observed in every direction of space—the lingering signature of creation. Massively parallel-processing computers, with cellular automata and other kinds of computer-based mathematics, may lead to the most sweeping transformation of science since the Scientific Revolution of the seventeenth century. Hot fusion and superconductivity may transform the way civilization produces and uses energy. The Genome Project will provide a complete genetic blueprint for a human being. The scanning tunneling microscope allows us to "see" individual atoms and move them about at will.

These things are important, hugely important. Yet most of us know little or nothing about them. Instead, we wallow blissfully in the wildest forms of superstition and discredited myths.

Why? The high school students with whom I spoke provided thoughtful answers. Pseudoscience and superstition have a human face, they said; science is remote and forbidding. Pseudoscience and superstition are easy to understand; science is complex and inaccessible. Magic and superstition are steeped in history; science is as new as yesterday. The popular media are awash in pseudoscience and superstition; science is relegated to a few programs on public television. For all of these reasons, so thoughtfully articulated by the students, True Belief is in ascendancy and Skepticism has its back to the wall.

Credulity Index

Are you a Skeptic or a True Believer? Try this test to measure your Credulity Index. If you are uncertain about a statement, but more open to the possibility than not, count it a True. If you are uncertain, but skeptical, count it a False.

	TRUE	FALSE
The Earth is less than 10,000 years old.	☐	☐
Extrasensory perception (ESP) is real.	☐	☐
Humans were created by God essentially as they are today.	☐	☐
Some of us lived earlier lives.	☐	☐
UFOs probably have an extraterrestrial origin.	☐	☐

	TRUE	FALSE
Miracles happen.	☐	☐
Our souls will spend eternity in heaven or hell.	☐	☐
Human consciousness exists independently of the material brain.	☐	☐
Certain religions are more favored by God than others.	☐	☐
Horoscopes are based on real influences of the stars.	☐	☐
God hears and sometimes answers our prayers.	☐	☐
It is worth looking for the Loch Ness monster.	☐	☐
A personal creative force guides the evolution of the universe.	☐	☐
The Egyptian pyramids may be the work of extraterrestrials.	☐	☐
The spirits of the dead are still with us.	☐	☐
The Bible (or Koran or other sacred text) is the word of God.	☐	☐
Scientists will never create life in the laboratory.	☐	☐

	TRUE	FALSE
Certain people have the power to predict the future.	☐	☐
Angels exist.	☐	☐
The Apocalypse is imminent, perhaps at the year 2000.	☐	☐

The sum of your "True" answers is your Credulity Index. Skeptics will have a low Credulity Index; True Believers will have a high Credulity Index.

What these statements have in common is that there is no scientific evidence to support them—that is, no reproducible empirical evidence that would pass muster in a peer-reviewed scientific journal. Some of these statements are accepted by the vast majority of us; others are believed by a small minority. If a 1995 Gallup poll is to be believed, about 80 percent of us believe miracles happen, and another 10 percent are open to the possibility. Three-quarters of us believe in angels. About a quarter of us believe in reincarnation and astrology, with another 20 percent open to the possibilities. Virtually all Americans (more than 95 percent) say they believe in God or a universal spirit.

It is the rare person who does not at some point in her life utter a prayer, call upon the assistance of a personal force greater than herself, or hope for a miracle. The most avowed Skeptic, faced with calamity, might harbor a hope that the laws of nature will admit exceptions. Skepticism offers little consolation in times of trial. When oblivion yawns before us, a high Credulity Index is the natural response. "Believe in me, and you shall live forever," offers the guru—an irresistible pitch.

Astrology

Some years ago, I wrote a column that was, I thought, a blistering debunking of astrology. It was about the time then-President Ronald Reagan and First Lady Nancy Reagan admitted to consulting astrologers on matters of state. Polls showed belief in astrology was on the rise. Hardly a newspaper or magazine in the country was without a horoscope (still true, alas). I pointed out the complete absence of any reproducible, empirical evidence linking individual human lives to the positions of celestial bodies. I described controlled tests of astrology, all of which had proved negative.[2] I stressed the positive virtues of scientific skepticism and suggested that astrology is fundamentally antirational. For most of my audience—people who read the science pages of the newspaper—I was preaching to the converted. For the rest, no amount of debunking would have made any difference.

One good thing came out of the column. I entered into an agreeable correspondence with a professional astrologer who was convinced of the validity of his craft, and who, as far as I could tell, was motivated by an unselfish desire to help others. He offered gentle, patient responses to my not-so-gentle criticisms. He sent me several of his best-selling books on astrology. They were lively, well written, and fun. As self-help books go, this particular author's works contained much good, sensible advice. They evoked a sense of wonder, a positive attitude toward people, and (paradoxically) a healthy sense of personal responsibility. However, I learned nothing from the books that convinced me that astrology is anything more than a silly, mostly benign superstition.

What I did learn is that there is no way for a scientist to convince an astrologer that his craft is without foundation, and no way an astrologer will convince a scientist otherwise. It is not so much a matter of evidence as *attitudes toward evidence*. The astrologer and the scientist have different criteria for truth—the one anecdotal and personal, the other empirical and institutional—and, consequently, little hope of resolving their differences. "I don't care what experiments *appear* to show; I just know from experience that it works," says the astrologer. When a scientist sees a person ordering his life by the stars, he sees the surrender of reason. When an astrologer sees a scientist "debunking" astrology, he sees bias or blinkered dogmatism.

One correspondent to the newspaper took issue with my column this way: "Raymo's argument against astrology is the usual one: Astrology can be done away with by simply declaring it irrational. In other words, if we cannot understand why it works, it must not work. The same flawed argument could be used against electromagnetism, particle physics, and the force of gravity, with equally senseless results." And it's true. I don't understand in any ultimate sense why electromagnetism, particle physics, or the law of gravity works. Nobody does. The point is, electromagnetism, particle physics, and the law of gravity *do* work, in a way that astrology does not.

Experiments of the most exquisite sensitivity can be devised to test the former theories, experiments that can be performed by Believers and Skeptics alike with identical results. Radio communication, nuclear power, and the space program are spectacular testaments to the fact that electromagnetism, particle physics, and the law of gravity work. On the other hand, as noted earlier, every scientifically reliable test of astrology has been negative. Whenever professional astrologers have been asked in blind, controlled experiments to match horoscopes with personality profiles or personal histories, their success rate has been no better than chance. Then why do astrologers and the people who read horoscopes continue to insist that astrology "works"?

Here's why. I once had my birth chart done by an astrologer. She labored long over ephemerides and graphs, then told me I was sensitive, intelligent, basically generous but sometimes self-indulgent, inclined toward optimism but subject to occasional bouts of depression. (Wow! Right on!) In other words, she told me just what I wanted to hear, in language so vague as to be untestable. In spite of reliance on numbers, graphs, and even computers, astrology has nothing to do with science.

Skeptics should ask themselves what is behind the incredible popular appeal of astrology. Until we can find a way to reconcile scientific skepticism with spirituality and religious feeling, there will continue to be a wide audience for pseudoscience and superstition. On the other hand, True Believers should ask themselves if they really want to live in a world that is ruled by slipshod attitudes toward evidence, where emotions rule intellect.

We are frequently reminded by astrologers that such great scientists as Kepler and Newton believed in astrology. True enough. But Kepler and Newton lived at the dawn of the scientific era. Kepler's mother was nearly burned as a witch, and Newton's university was closed because of an outbreak of the Black Death. Witchcraft flourished because people imagined causal connections where none existed, on the basis of anecdotal evidence, and the plague vanished from Europe when people took note of causal connections that could be verified empirically and reproducibly. It is no coincidence that witchcraft and plague vanished from the Western world at about the same time that astrology was finally discarded from science.

Prayer

Astrology has much in common with the cosmological systems of traditional religion. It is human-centered. It is based on miracles (the assumption of a causal agent that lies beyond the bounds of orthodox science). It affirms a link between the cosmos and our personal lives. It is fundamentally optimistic. (How often have you read a discouraging horoscope?) It is embedded in an impressive matrix of lore, ritual, and historical precedent. The only important difference between astrology and mainstream religious faiths is the lesser number of people who profess belief in astrology. The supernaturalist cosmology of world religions is so deeply grounded in our history, so intimately linked to our traditional understanding of ourselves, so thoroughly meshed with belief in an afterlife, that it is hard to imagine what life would be like without it. Yet there remains the troubling inconsistency between our way of knowing (science) and our way of believing (religion).

This inconsistency can be illustrated by our attitudes toward healing. According to a *Time* magazine survey, 77 percent of Americans believe God sometimes intervenes to cure people who have a serious illness, and 73 percent believe that praying for someone else can help effect a cure.[3] Even Skeptics will admit that prayer might sometimes ameliorate illness by harnessing the brain's influence over the body's physiological processes or by promoting a relaxation response. But can prayer addressed to God on a patient's behalf—*without the patient's knowledge*—heal?

In 1988, Randolph Byrd published results of a study at the San Francisco General Medical Center that has been widely quoted as offering scientific evidence for the medical benefits of intercessory prayer.[4] Over a ten-month period, Byrd randomly divided nearly 400 patients in a coronary care unit into two groups. One group was prayed for by born-again Christians outside the hospital; the control group received no assigned prayer. Neither physicians nor patients knew which group the patients had been assigned to. According to Byrd, patients receiving intercessory prayer required less ventilatory therapy and fewer antibiotics and diuretics than the control group. His tentative conclusion: Prayer works.

However, no significant differences were noted in such variables as length of hospital stay or mortality. Furthermore, Byrd's study has been faulted on statistical and procedural counts, even by some believers in the power of intercessory prayer. Also, given the impossibility of knowing who has actually been prayed for, by whom, and how often, it is hard to imagine how any scientific study of intercessory prayer can be conclusive one way or the other.

These reservations are seldom mentioned when the results of Byrd's study are quoted, as, for example, in a 1996 *Time* magazine cover story on alternative healing strategies.[5] Nor do we hear much about two earlier medical studies that failed to provide *any* statistical support for the success of intercessory prayer.[6] Of course, negative results prove nothing to Believers; after all, God may simply refuse to cooperate with scientific tests of his power. If Byrd's ostensibly positive results were somehow confirmed in an unambiguous way, it would represent a stunning challenge to the scientific worldview. Suffice it to say that the medical community is highly skeptical.

But not universally so. Larry Dossey is a formerly orthodox physician and successful author who has made a name for himself as a proponent of alternative healing strategies, including prayer. He has expressed reservations about Byrd's study but is open to the healing power of prayer, even if the patient doesn't know she is being prayed for. He argues for a new kind of medicine that takes account of a "collective mind," a mind not localized to brain or body.

The respected Harvard cardiologist Herbert Benson is another traditionally trained physician who is pushing for the scientific study of alternative therapies. He also recognizes flaws in Byrd's study but thinks intercessory prayer is something that should be studied in a rigorous scientific way. With his colleagues, he has designed a study of intercessory prayer that he believes meets the strictest scientific standards. Like any good scientist, Benson is interested in the causal mechanisms at work in alternative therapies, including prayer. And certainly, the mind-body connection is pretty much unexplored territory in science.

Nevertheless, skepticism remains high among scientists. In a review of Benson's book *Timeless Healing: The Power and Biology of Belief*, biologist Irwin Tessman and his physicist brother Jack Tessman take Benson to task for what they consider misleading use of evidence. "It is undeniable that the mind affects the body in many ways," write Tessman and Tessman. "Therein lies a fertile field for rigorous science; also a fertile field for exaggerated claims, uncontrolled studies, flawed statistics, mind-boggling illusions, and anecdotal reports."[7]

Benson, Dossey, and other alternative-therapy physicians can be admired for their efforts to bring mind-body interaction within the fold of rigorous science. And, certainly, anything that leads to healthier lifestyles and less dependence upon interventional medicine is to be commended. But the new alternative-medicine gurus should be careful to acknowledge what Dr. Gerald Weissmann writes in *Democracy and DNA: American Dreams and Medical Progress*: "There is no homeopathic, ayurvedic or New Age practice that can prevent pandemics of plague, protect the earth from decay or pollution, or prolong the life of a tot with a congenital hole in its heart." (He might have included intercessory prayer among alternative therapies with scant nonanecdotal evidence for success.)

In the fall of 1976, the first outbreak of Ebola virus appeared at a remote hospital in Yambuku, Zaire, staffed by Belgian nuns. Within a few weeks the virus took a deadly toll, causing horrible deaths from internal hemorrhaging. The courageous nuns did what they could with their limited pharmaceuticals and scientific training, but the plague raged unabated. When the first representatives of the Centers for Disease Control (CDC) in Atlanta and the World Health Organization (WHO) arrived at Yambuku, they found the hospital pitifully cordoned off with a strip of gauze bandage, the surviving nuns reduced to prayer.

The scientists set to work, taking blood and tissue samples for study in the field or for shipping to Atlanta for more detailed analysis, working out transmission patterns of the disease, and looking for the animal or animals that might be reservoirs or vectors for the virus. They discovered that the virus was a new and virulent strain, spread initially by the reuse of scarce syringes at the hospital. Through a combination of quarantine and strict hygiene, the epidemic was brought under control.

Meanwhile, a similar viral epidemic was ravaging remote villages in southern Sudan. Again, the international disease-control scientists made their way to the scene. Again they discovered that unhygienic hospital practices were instrumental in spreading the disease, compounded by tribal methods of preparing the bodies of victims for burial. (The kin of the deceased used their bare hands to remove undigested food and feces from the dead person's internal organs.) When the hospital was closed and the funereal cleansings stopped, the epidemic abated. It was not prayer that contained this pandemic-threatening virus, but science.

Thwarting the rapid spread of deadly new pathogens in a world increasingly linked by rapid mass movements of people will require permanent vigilance on the part of health professionals. Our lives literally depend upon them. But most of us have not the foggiest knowledge of what these courageous medical personnel are up to, or how they put themselves on the front lines of the war against disease.[8]

Meanwhile, books by New Age healing gurus fly out of the bookstores. We *know* that medical science is the best hope for healing and health, but we *believe* as if miracles are possible. Emotional states have an undoubted impact upon physical health, and it is certainly true that the medical establishment is often woefully unsympathetic to our emotional needs, but—as the earnestly praying Belgian nuns in Yambuku experienced to their distress—God has no role in the micromanagement of viruses and bacteria.

Our best hope of negotiating our precarious future will be, as Gerald Weissmann urges, "explaining facts by facts." Alternative healing therapies deserve open-minded study, but someone, somewhere, had better retain a low Credulity Index or we are in danger of pandemic chaos.

SIX

Close Encounters of the Improbable Kind

Science has beauty, power, and majesty that can provide spiritual as well as practical fulfillment. But superstition and pseudoscience keep getting in the way, providing easy answers, casually pressing our awe buttons, and cheapening the experience.

Carl Sagan

SINCE THE 1940S, UFOs have been a favorite folly of the True Believer, second only to the religious fundamentalist's affection for a literal Adam and Eve. I was once a UFO buff myself. As a teenager in the late 1940s and early 1950s, I read everything written on UFOs. That was the golden age of UFOlogy, and I was convinced that the breathless reports of flying saucers were onto something. The evidence, although anecdotal, seemed terribly impressive. The fact that UFOs weren't acknowledged by science could be put down to a government cover-up. Somewhere in a secret hangar on an air force base in Ohio the evidence was piling up: the carefully investigated eyewitness reports, the photographs and radar records, the crashed saucers, maybe even the bodies of aliens. It was only a matter of time, I thought, until the story broke into the open and close-minded scientists recognized the phenomenon for what it was.

Then I went off to college, studied science, and developed some skills of critical thinking. I learned how to evaluate evidence—how to apply the keen edge of Ockham's razor. And when I went back to the breathless books about UFOs, I found nothing that could stand up to scrutiny. What I found instead were delightful stories, spiced with a delicious dollop of fear, of superior beings from somewhere in the heavens who took a special interest in humans. What I found, in other words, was wishful thinking—angels in a new guise.

Forty years later, flying saucers are still with us, more stubbornly pervasive than ever. What is more, they have taken to abducting us by the millions. And whatever it is they are interested in, it has (surprise, surprise) something to do with sex.

According to a 1991 survey of 6,000 adult Americans by the Roper Organization, approximately 1 out of every 50 adult Americans has experienced at least four of the five characteristics of a typical UFO abduction:

- waking paralyzed with the sense of a strange figure or presence in the room;
- experiencing an hour or more of "missing time";
- having a feeling of actually flying through the air without knowing how or why;
- seeing unusual lights or balls of light in a room without understanding what has caused them; and
- discovering puzzling scars on one's body, without remembering how or why they were acquired.

The Fund for UFO Research interprets these results to mean that upward of 3.7 million Americans may have been abducted by extraterrestrials. Three-point-seven million! I'm beginning to feel a bit left out. You'd think I'd be the sort of fellow the aliens would go for. As an enthusiastic stargazer, I spend lots of time under the night sky. Almost every clear night I'm out there scanning the skies. I've even seen a few strange lights in the sky I can't explain. But I've never been abducted. Why are they avoiding me?

In the early 1990s, the alien abduction phenomenon was given a mighty boost by the endorsement of Dr. John Mack, professor of psychiatry at the Harvard Medical School and Pulitzer Prize–winning biographer of T. E. Lawrence. Mack believes that hundreds of thousands of American men, women, and children may have experienced UFO abductions, or abduction-related phenomena. He bases his belief on more than 100 interviews, often involving hypnosis, with men and women who "remember" having been taken aboard spacecraft for bizarre sexual experiments, involving sperm taking, artificial insemination, removal of embryos, surgical implantation of "tagging" devices, and probes of body cavities. Mack is one of many observers who think an alien race may be interbreeding with humans or tampering with our genes. Presumably, the purpose of these procedures is to create a hybrid race of half-humans, half-aliens who will save our hopelessly flawed planet Earth from self-destruction.

What is offered by the so-called alien abduction phenomenon is no less than a vision of planetary redemption, and nothing could be more welcome by UFOlogists than the support of a scientist with Mack's impressive credentials. Here is benediction from on high, the establishment seal of approval—a real honest-to-God Pulitzer Prize winner and, best of all, a Harvard professor. Suddenly, the value of abduction stock shoots up. When science offers skepticism, True Believers hurl epithets of "close-mindedness" and "tunnel vision." But let a credentialed scientist appear to confirm their True Belief, and they are quick to wrap themselves in the mantle of science's trusted authority—or in whatever threads of that mantle they are able to collect. As these things go, Mack is a very big thread indeed.

Not even Harvard professors, it seems, are immune to our need for redemption (or at least attention) by creatures from the sky. Mack's benediction comes with an interesting twist. For almost half a century, True Believers have insisted that UFOs are of extraterrestrial origin and that humans have been forcibly taken aboard spaceships for physical examinations or sexual encounters, and Skeptics have just as vigorously dismissed such reports as frauds, delusions, or mass hysteria. What is new in recent years is the appearance of a third slant to the debate, typified by books such as Keith Thompson's *Angels and Aliens: UFO's and the Mythic Imagination*, David Jacob's *Secret Life: Firsthand Accounts of UFO Abductions*, and Mack's own *Abduction: Human Encounters with Aliens.*[1] People in this camp avoid the stark polarities of the traditional debate: physical reality versus delusion, fraud, or wishful thinking. They reject, in Mack's words, "boundaries between the material and the psychological, the mythic and the real, as well as distinctions between symbolic and literal, and even . . . the polarities of true versus hoax."

It's difficult to get one's teeth into this sort of thing. If a person testifies (often under hypnosis) that she was removed from her home, subjected to physical examination on a stainless-steel table aboard an alien craft, and perhaps tampered with sexually and/or genetically, then I'm inclined to say either it happened or it didn't. Abduction researchers have a more slippery notion of truth. Something is going on, they say, that we can't hope to understand, involving technologies and realities beyond our wildest imaginings. If the abductors can travel to Earth from another planet, pass through walls, and render their victims invisible, then who are we to insist upon the knowledge categories of merely human science?

Why should Skeptics believe this apparent nonsense? Well, because of the consistency of abduction reports, say True Believers—most victims describe remarkably similar experiences—and because many abductees have marks and scars on their bodies of unknown origin. This kind of evidence provokes a weary sense of déjà vu, suggesting another explanation for the so-called abduction experiences, something less dramatic and more familiar than little gray bug-eyed, sex-obsessed men in spaceships.

Consider the striking parallels between the alien

abduction phenomenon and the witchcraft hysteria of the late Middle Ages:

- The investigators of the phenomenon are mostly male.
- The victims are mostly female, and mostly young.
- The victims are often wakened in the night with a sense of strange presences in the room. They are taken away through the ceiling and participate in experiences of a sexual nature.
- The testimony of the victims about their experiences shows remarkable consistency. Handbooks are written describing typical experiences. The consistency of the testimony is taken as evidence that the phenomenon is objectively real.
- Victims are sometimes afflicted with puzzling scars, injuries, or pains.
- The physical characteristics of the abductors, their presumed place of origin, and their mode of transport are consistent with the popular imagination of the time.
- The investigators participate in eliciting testimony from the victims (torture then; hypnosis now).

Today, few of us believe in witches, or in nighttime commerce with devils. We are inclined to view the witch craze of earlier centuries as an aberrant psychosocial phenomenon, brought to a timely end by the Scientific Revolution and the Enlightenment. However, the lesson of the parallels should be obvious: The so-called alien abduction phenomenon is telling us more about ourselves than about bug-eyed humanoids in flying saucers. There is no need to suppose interstellar travel or even new categories of truth and falsehood; the cause of the abduction reports lies closer to home—in the mysterious religiosexual complexities of the human psyche. The tragic mass suicides of the Heaven's Gate cult in 1996 are an example of what can happen when extraterrestrial nonsense becomes cloaked with a mantle of credibility, in this case a supposed photograph by a credentialed astronomer showing an alien spacecraft following Comet Hale-Bopp. The photograph was later shown to be spurious.

Coincidence and Evidence

In mid-1992, I wrote a *Globe* column about alien abductions, in which I pointed out the many striking similarities between contemporary abduction reports and the witchcraft hysteria of the late Middle Ages. When the column appeared, John Mack telephoned and we had a chat. Let me say at once, he was fun to talk to—thoughtful, intelligent, and sincere. I liked him immediately and don't doubt for a moment his earnest conviction that he is helping his patients by taking seriously their reports of alien encounters. Of course, neither of us convinced the other of the correctness of our views.

Some weeks after our chat, a friend gave me an audiotape of one of Mack's public lectures in which he referred to our telephone conversation, then added: "Finally, in exasperation, I said to him, 'Look, Chet. A UFO could land on Boston Common. Channel 5, Channel 7, and Channel 4 could all have films on the nightly news to show us. The *Boston Globe*, the *Boston Herald* could have big articles about it, and you still wouldn't believe, would you?' And Chet said, 'No, I wouldn't.' " Big laugh from Mack's audience.

I don't remember the details of my telephone conversation with Mack, but I suspect that when recounting our conversation he exaggerated a bit for dramatic effect. That's OK. I don't mind being the butt of the joke, and what he said is a fair representation of my views. Further, I wouldn't want to fall into what biologist Richard Dawkins calls "the Argument from Personal Incredulity": If it seems impossible to me, it must not be true. Anything, even abduction by aliens, is possible.

However, the arrival of a UFO from space would

be an event so far beyond our normal experience that any sensible person should ask for compelling, irrefutable evidence. After all, there are other possible explanations for TV and newspaper reports of a Boston Common landing: an April Fools' joke, a hoax, an episode of mass hysteria. I'd want to go to the Common and see the ship with my own eyes. Or, failing that, I would want reliably documented eyewitness testimony from Skeptics, not elicited under hypnosis. I would want material evidence: close-up photographs taken by credentialed news photographers, spaceship artifacts, a few of those "tagging" devices removed from the bodies of abductees by Skeptical doctors, and so on. As Carl Sagan said, extraordinary claims require extraordinary evidence.

John Mack believes he has extraordinary evidence in his interviews with supposed abductees. But he fails to follow one of the bedrock principles of science: *Don't make your explanations more complicated than what is absolutely required to explain the phenomenon.* Obviously, any phenomenon can have any number of explanations. A Skeptic will favor the explanation that involves the least number of causes that take us beyond what we already know to be true. For example, I've been missing a grass scythe from the toolshed of my house in Ireland for ten years, and you will recall from an earlier chapter that I live on "The Fairies' Road." Shall I assume the fairies took it? Only if a simpler, less tenuous explanation will not suffice. I've been known to be absentminded. Someday I will probably find the tool in the tall grass where I absentmindedly left it. But if I believed in fairies to start with, then no doubt I'd take the missing tool to be confirming evidence for my belief. *Coincidence is the evidence of the True Believer.*

By contrast, Isaac Newton put the Skeptic's faith this way: "We are to admit no more causes of natural things than such as are both true and sufficient to explain their appearance." Einstein said: "The grand aim of science . . . is to cover the greatest possible number of empirical facts by logical deductions from the smallest possible number of hypotheses or axioms." Ockham's razor, reexpressed by Newton and Einstein, is the defining difference between Skeptic and True Believer.

According to a 1996 Gallup poll, more than half of Americans are willing to entertain an extraterrestrial origin for UFOs. The number hardly matters; the search for scientific truth is not a matter of popular vote. Most scientists believe there is a more economical way of explaining "alien abductions" than by invoking the literal kidnapping onto spaceships of millions of humans—namely, psychological explanations of one sort or the other. We *know* humans hallucinate, repress memories, self-delude, lie. We *know* humans have sexual fantasies and hang-ups. We *know* humans sometimes wake up in the dark disoriented and confused, with a sense of a strange presence in the room. We have seen the pattern before, in the witchcraft hysteria of the late Middle Ages. The subjects of all thirteen chapter-length case studies in Mack's book (selected from the forty-nine patients he studied most carefully) exhibit widely recognized characteristics of fantasy-prone personalities, based on internal evidence of the book.[2]

Of course, John Mack rejects the possibility that he is dealing with fantasy. In his recorded lecture, he suggests that the reason most scientists, such as myself, are unsympathetic to his evidence is that we are not open to explanations that run counter to conventional wisdom. He's right. We should "run counter" only when we need to. That's the strength of the Skeptic's way of knowing. Scientific knowledge is anchored to reality by robust mooring lines.

This story has an amusing epilogue. I ended my *Globe* column on John Mack and alien abductions with this statement: "Tell you what, John. Pass the word through your abductee contacts. I'll be waiting on the college quad at midnight a week from tonight. I volunteer myself for alien abduction experiments. I doubt if anyone will show up to spirit me away—but I'm prepared to be astonished." The result was predictable, and I should have anticipated it. Almost immediately after the column appeared, signs went up all over my college campus: "See

Raymo abducted by aliens! On the quad. Midnight, Monday, April 18."

On the appointed night, 1,000 high-spirited, spring-fevered students gathered on the quad, many of them in alien costume (consuming dozens of rolls of aluminum foil). A landing area had been roped off. The college radio station had set up a special booth and was blaring *Star Wars* theme music into space. Reporters from the local newspapers had been alerted and were on hand. I couldn't stay at home. So I joined the throng, with an overnight bag containing a voluminous (ersatz) sperm sample and a fresh change of underwear. At midnight, 1,000 voices, many of them cheerily inebriated, counted down the moment of truth. A rousing good time was had by all. I wish John Mack had been there; I think he would have had a good time too. The aliens didn't show.

Mathematical Immortality

Not long after John Mack's book appeared, another highly credentialed scientist upped the ante in the redemption sweepstakes and bestowed upon True Belief another welcome benediction from on high. Respected physicist Frank J. Tipler published an extraordinary book called *The Physics of Immortality: Mod-*

ern Cosmology, God, and the Resurrection of the Dead.[3] The book caused quite a stir, not least, I suspect, because a respectable scientist was telling us what we wanted to hear. The book played perfectly into our love-hate relationship with science. As long as science suggests that our lives are brief and inconsequential in the cosmic scheme of things, we dismiss it as flawed and close-minded. But let science appear to suggest that we are of cosmic importance and, best of all, immortal—well then, at last scientists are getting it right.

An ad for Tipler's book in the *New York Times Book Review* gushed: "Renowned physicist Frank J. Tipler has found what surely must be the most dramatic scientific proof of all time. His mathematical model of the end of the universe, the Omega Point Theory, leads to the stunning conclusion that God exists, that there is a heaven, and that there will be a resurrection. In easy-to-follow layman's language he explains the theory and its historic consequences for our lives, our systems of belief, and our world." And, as if that were not enough, the ad continued: "The clash between science and religion began with Galileo. It ends here." Wow! The most dramatic scientific proof of all time. God, heaven, and resurrection proved by physics. You've gotta be kidding! It turns out that Doubleday's advertising department can't hold a candle to Tipler's own overheated hyperbole. He writes in the introduction to his book: "If any reader has lost a loved one, or is afraid of death, modern physics says: 'Be comforted, you and they shall live again.' "

Tipler is a professor of mathematical physics at Tulane University, erudite, broadly knowledgeable, and highly intelligent. His book is replete with references to writers as various as Heidegger, Aquinas, and St. Paul. The long scientific appendix is chockablock with complex mathematical equations. Tipler clearly takes what he's doing seriously. He assumes (as do most scientists) that the universe began with a Big Bang from a mathematical singularity that contained all that exists today in a state of pure energy at infinitely high temperature. The universe is presently expanding from the primeval impetus. Tipler then assumes that the mass density of the universe is such that the expansion will slow down and give way to contraction. (There is no convincing observational evidence that this is the case.) If so, then hundreds of billions of years from now the universe will end in another singularity, sometimes called the Big Crunch, but called the Omega Point by Tipler, in language reminiscent of the Jesuit philosopher-scientist Pierre Teilhard de Chardin.

Tipler defines life as information processing. The "self," he says, is an enormously complex computer program running on a biological computer. In the final collapse of the universe to the Omega Point, all of the programs of every person who ever existed (and who ever could have existed) will be mathematically re-created in the mind of an all-seeing, all-loving God that Tipler identifies with the universe itself. All of this is ostensibly derived by Tipler from the laws of quantum physics. That's the gist of his ideas, although my brief description hardly does justice to the mathematical subtleties. If you can find in such a theory the consolations of religion, then you are welcome to them.

The Omega Point theory is "science," Tipler claims, arrived at in exactly the same way as physicists calculate the properties of the electron. Indeed, with the Omega Point, he says, religion becomes a part of physics: an experimentally verifiable—or falsifiable—proof of a personal, omniscient, omnipresent, all-powerful God, and of the resurrection of every human to live again in bliss. What is on offer here is not merely a brief fling in a spaceship, but everlasting life; not an out-of-body experience, but eternal happiness in heaven and you can take the body with you. According to Tipler, the Omega Point theory answers the biggest questions asked by the human mind: Who am I? Why am I here? What will be my ultimate fate?

He's wrong. The biggest questions that come to my mind are why did Tipler write this book, why did Doubleday publish it, and why would anyone shell out $24.95 to buy it? First of all, don't be fooled by that "easy-to-follow layman's language" claim.

Anyone would need a sophisticated grasp of science to understand the physics behind the theory. Also keep in mind that Tipler does not even believe his theory himself, at least not yet. He tells us, on page 305, that so far there is no evidence in its favor but theoretical beauty. Yet the "theoretical beauty" of a highly speculative and impossibly abstract mathematical theory is supposed to console us upon the death of a loved one? Pure messianic delusion.

Tipler also offers his own theory on sex in the afterlife, conveniently reducing his argument to absurdity. "My students—mainly young unmarried males—often ask me, 'Will there be sex in heaven?' " Tipler writes. He answers "yes." Indeed, he says, the all-powerful Omega Point will be able to match every young unmarried male with the most beautiful woman he has ever known—no, with the most beautiful woman who is logically possible. (This is infinitely better than the sexual attentions of aliens; this is the real thing, an eternal pie-in-the-sky celestial orgasm with the partner of your dreams.) Tipler wonders if the nervous system of the resurrected individual will be able to stand so much beauty, and (with a mathematical wave of the hand) answers in the affirmative. The only disadvantage to this promise of resurrected bliss is that you will have to wait hundreds of billions of years after your death for the crunching universe to re-create your quantum wave functions.

If the possibility of requited lust in the Big Crunch afterlife sounds good, then for $24.95 you can have the whole eschatological enchilada. As for myself, if I wanted the consolations of an afterlife—with or without sex—I could find a more consoling myth in any church on the planet. Is Tipler's theory the end of the clash between science and religion? Hardly. Is it the most dramatic scientific proof of all time? Give me a break. This is the kind of pseudophysical mysticism that gives physics a bad name.

Marketing Pseudoscience

Mack and Tipler offer more under the banner of scientific respectability than their empirical evidence can support, and by doing so they dangerously blur the boundary between science and pseudoscience. They may be the best-credentialed names currently offering benediction to the paranormal, but they are not alone. Bookstore shelves burgeon with books by fringe scientists and pseudoscientists purporting to offer what I'll call New Theories of the Universe (NTUs). Herewith, rules for packaging a NTU if you want to offer one of your own:

1. Give your NTU an aura of real science. Use words like *cosmic, morphic, plasma, energy matrix, astral, etheric, resonance.* By all means invoke the Heisen-

berg Uncertainty Principle, which can be interpreted by the uncritical mind to mean "anything goes." Quantum theory and chaos theory can be made to sound vaguely compatible with the paranormal.

2. Flaunt your credentials. Put an M.S. or Ph.D. after your name on the jacket of your book; it doesn't matter in what field of study you acquired your degree.

3. Make sure your NTU is easy to understand ("easy-to-follow layman's language"). You may use schematic drawings of warped space-time, but, please, no mathematics. If you *must* use mathematics, keep it to an impressive but unreadable "technical appendix."

4. Don't hesitate to point out all the things that real science can't explain: the origin of life, the development of organisms, consciousness, dreams.

5. Remember, your NTU needs evidence. A good rule of thumb: You can always track down at least a dozen purported occurrences of any phenomenon.

6. Distance yourself from the most simplistic superstitions. For example, make fun of newspaper horoscopes and the stories in supermarket

tabloids. But also make sure that your NTU is vague enough to allow for—or at least not prohibit—astrology, aliens, ESP, psychokinesis, ghosts, immortality, and other popular paranormal phenomena.

7. Keep your NTU human-centered. Real science tends to make people feel isolated, forgotten, like cogs in a machine. A good NTU makes every individual feel like the center of a cosmic web of influences.

8. Include a bit of sex.

9. Don't be afraid to evoke the wrath of the scientific establishment; this will prove you are onto something big. For example, the best thing that ever happened to Rupert Sheldrake was a bit of intemperate editorializing in the science journal *Nature.* Sheldrake (an ex–Frank Knox Fellow at Harvard, Ph.D. in biochemistry from Cambridge University) is the author of *A New Science of Life: The Hypothesis of Formative Causation* and *The Presence of the Past: Morphic Resonance and the Memory of Nature*, books claiming that everything from a crystal to a human becomes what it is because it remembers what it is supposed to be. When the former book was published in 1981, the editors of *Nature* called it an "infuriating book . . . the best candidate for burning there has been in many years,"

and immediately propelled the book onto the best-seller lists. Subsequent editions of Sheldrake's book have used the *Nature* denunciation as a publicity blurb. Why is Sheldrake's theory infuriating? Why is the dreaded scientific establishment running scared? Could it be that . . . ?

10. Remember those famous lines from *Hamlet:* "There are more things in heaven and Earth, Horatio, than are dreamt of in your philosophy." Scientists don't know everything. The key to success for any good NTU is to amass enough anomalies, coincidences, oddities, exceptions, prodigies, and wonders that the sheer bulk of your data will convince the reader that your theory is correct. After all, if orthodox science can't explain All of This, then alien abductions, the Omega Point, or (*insert your own theory*) begins to look better and better.

Hard Truth or Easy High

Not long ago, I was invited to a small midwestern college to talk with a group of talented young nature writers. They had read a couple of my books and generally approved of the way I tried to relate science to human values. However, they took me to task for what they perceived as condescension on my part toward astrology, channeling, alien abductions, extrasensory perception, quantum immortality, and other New Age enthusiasms. In one of my books I had called these things "baloney." "Don't be so cocksure," they said. "Remember that even Galileo was a victim of close-mindedness." Their point is well taken. There is more to the world than meets the eye, and it behooves us not to dismiss anything too quickly. Such things as UFOs or ESP can't be ruled out a priori. However, I do not dismiss them quickly. The purported evidence has been subjected to exhaustive scrutiny; it is unconvincing. UFOs, ESP, and other paranormal phenomena do not bear up to the fierce competition of ideas that is the Skeptic's criterion for truth.

The students who took me to task were not unsympathetic to the scientific way of knowing, but they have not found much in science that answers their need to feel at home in the universe. What they are looking for, it seems to me, is a sense of stewardship for the Earth. They want to feel part of an organic system in which their individual existences make a difference. They want a reason to live, and they want a reason to live unselfishly. They want, in other words, all of those things that only religion can provide. Nevertheless, many of them have turned away from those traditional religions that they perceive to be in manifest conflict with science. They embrace instead New Age faiths that flourish on the margins of science—supernaturalism with a patina of scientific spiff. They align themselves with imagined powers and spirits that they perceive to be consistent with science, but which escape the grim uncertainties of the Skeptics.

"Give me the evidence for your belief," I asked them. "It makes me feel good," they said, or, quoting TV's *The X-Files*, "The truth is out there." They have become, in short, True Believers by default. And fair enough. Certainly, it is better to feel good than to feel bad. But the price the True Believer pays for feeling good can be a chasm between intellect and intuition, and exile from a scientific story of the universe that, like it or not, is *our* best story, a story that is empirically reliable and therefore more ultimately meaningful than any mishmash of New Age enthusiasms. The choice, on the face of it, is between a hard truth or an easy high.

SEVEN

The New Story of Creation

All great supernatural belief systems—indeed all philosophical systems, up till now—have catered to two central [spiritual] needs: the need for a rational understanding of the surrounding world, and the need for emotional security within it.

Nicholas Humphrey, psychologist

IN THE SPRING OF 1925, the Tennessee state legislature passed a law forbidding the teaching in public schools of "any theory which denies the story of the Divine creation of man as taught in the Bible," and teaching instead "that man is descended from a lower form of animals." John Scopes, a young biology teacher in the Dayton high school, defied the law by teaching Darwinian evolution and was brought to trial. Williams Jennings Bryan, eminent lawyer-politician and three-time candidate for president of the United States, assisted the prosecution. Scopes was defended by Clarence Darrow, the country's best-known criminal attorney, who had achieved notoriety defending Leopold and Loeb in a sensational Chicago murder trial. With these two titans pitted against each other, the country's attention was arrested. The sleepy town of Dayton, Tennessee, was temporarily transformed into the center of the universe.

Some Americans viewed the Scopes trial as a confrontation of reason versus superstition, enlightenment versus obscurantism, scientific skepticism versus blind commitment to religious dogma. Others saw the trial as a clash of theism versus atheism, morality versus immorality, angel-man versus monkey-man. What the trial drove home for everyone was the apparent antagonism of science and religion. In an unusual move, Bryan took the stand as an expert on the Bible. Darrow's dramatic cross-examination revealed Bryan's complete lack of scientific knowledge, rousing nationwide ridicule. Nevertheless, Scopes was found guilty and fined $100. Bryan died five days later. In 1927, the Tennessee state appellate court overturned the verdict. One would have thought that was the end of it. But the issue refuses to die. As recently as 1996, Tennessee legislators, urged on by Christian fundamentalists, again debated a law that would make it a criminal offense to teach evolution as "fact."

A 1993 Gallup poll posed these alternatives to Americans and asked which most closely represents their belief: (1) Humans developed over millions of years from less advanced forms of life, but God guided this process; (2) Humans developed over millions of years, but God had no part in the process; or (3) God created humans pretty much in their present form at some time within the last 10,000 years. Forty-six percent of Americans agreed with option 1 or 2. Forty-seven percent chose option 3. Seven percent had no opinion. It would appear that the country is about evenly divided between those who believe the Earth is millions of years old (actually billions), as scientists say it is, and those who believe the Earth is less than 10,000 years old, as implied by a literal interpretation of the Bible. This dramatic difference of opinion is worth examining more closely.

A Ball of Yarn

In an earth science course that I teach at Stonehill College, I ask students to make a time line of Earth history. One imaginative young woman returned to class with a melon-size ball of yarn. Each foot of yarn represented 10 million years. Major geologic eras—Precambrian, Paleozoic, Mesozoic, Cenozoic—were different colors of yarn, and multicolored buttons tied along the strand were keyed to important events in Earth history, described on an accompanying sheet of paper. There was not enough space to unwind the ball in the classroom, so we took it outside into the college quad. When the yarn was unrolled, it was 450 feet long, the length of one and a half football fields, representing 4.5 billion years of Earth history. I stood with the class at one end of the strand—the present—and looked off to the other end where a student stood at the epoch of the Earth's beginning, solitary and distant. It was an impressive demonstration of the abyss of geologic time. On this same scale, the 10,000-year age of the Earth according to nearly half of Americans *is about the thickness of a piece of paper.*

When I saw the ball of yarn rolled out, it occurred to me that there is probably no issue upon which Americans are more decisively divided than the age of the Earth. Neither Democrats and Republicans, nor internationalists and isolationists, nor environmentalists and big business, nor even advocates of pro-life and pro-choice are so starkly opposed in their opinions. For most contentious issues in American life, there are shades of gray, a spectrum of opinion, extremists and moderates. But not for the age of the Earth. *One and a half football fields versus the thickness of a piece of paper*. Nothing in between. It's either the literally interpreted Bible's 10,000 years (more or less) or science's 4.5 billion, depending upon which authority one accepts. This is more than a quibble about the interpretation of evidence; it is a fault line in our culture.

I know of not a single article in the vast body of international, peer-reviewed scientific literature offering evidence for a recent creation. On the other hand, library shelves groan under evidence for the evolution of life over hundreds of millions of years. The scientific evidence for a geologically ancient Earth is overwhelming, yet it is rejected by nearly half of Americans. What is the reason for this extraordinary denial of what every scientist knows to be true? Why do so many people latch onto biblical creationism with such fierce and unbreachable faith?

The chasm is not simply one of religion versus science. Three-quarters of the people who opted in the Gallup poll for a geologically ancient Earth believe God guides the process of evolution. More than 95 percent of Americans profess belief in some sort of superior being, and this presumably includes most evolutionary geologists. Nor is the divergence of belief regarding the age of the Earth strictly one of faith versus reason. Many people who believe in a geologically ancient Earth do so because they trust the reliability of the scientific way of knowing, even if they do not fully understand all of the arguments and evidence. And some believers in a young Earth have elaborate rationales for their opinion.

Here we see how stark is the choice between skepticism and True Belief—one and a half football fields versus the thickness of a piece of paper. It is not a choice that will be made by weighing up empirical evidence—say, 55 percent for a 4.5-billion-year-old Earth versus 45 percent for a 10,000-year-old Earth. No, the scientific evidence is more certain than that. It is 100 percent versus 0 percent. Astronomy, geology, chemistry, physics, and biology converge on the same conclusion: The Earth is geologically ancient.

A New Story

Science will not go away. What is known cannot be unknown. Knowledge that provides unprecedented power over nature will not be forsworn; supernaturalism, magic, and paranormalism cannot sustain a technological civilization. Science *works*, and it works so unexpectedly well that we must suppose that its success is no fluke. The Earth is 4.5 billion years old, or it is 10,000 years old; we must make the choice. Unless we are willing and able to make judgments like this, then knowledge has no meaning. Can we have our cake and eat it too? Science for our computers, mobile phones, and antibiotics; supernaturalism to make us feel good? Nearly half of us want it both ways.

Scientific skepticism is a half-empty bag. Science provides a reliable rational understanding of the natural world, but it does not address our need for emotional security. We stand at the end of that long, long strand of yarn and feel lost in an inhuman span of time. I stood there in the college quad with my students and held a piece of paper between my forefinger and thumb: "This," I said, "is all of recorded human history." I sensed a frisson of fear in my audience. I felt it myself. The universe of the geological eons *is* terrifying, like the space of the galaxies. Our lives are like a drop of dye in the sea, infinitely diluted. No wonder so many of us deny the evidence of our senses and turn to True Belief, opting for the security blanket, the thumb, the parent's embrace.

And the rest of us, the Skeptics, what do we hold on to? There is an irreducible intuitive core to our knowledge that is immune to organized skepticism: our sense of self, our sense of other, and our sense of the inexhaustible mystery of the world. We can no more deny these intuitions than we can deny the sensations of sweet or bitter or blue. They are givens. Nothing science says can make these intuitions go away or undermine their veracity. They are the bedrock of religious experience, upon which Skeptics will commence their search for emotional security.

Intellectuals have a tendency to blithely dismiss religious feeling, forgetting that the religious impulse is basic and universal to the human experience. The sense of self, the sense of other, and the sense of unfathomable mystery have been with us since the first spark of consciousness ignited a human mind. They remain the open door to the divine. We got ourselves into our present muddle by naming the unknown sources of these intuitions, then confusing the names with external reality. The cosmological underpinnings of much contemporary religious faith—animism, anthropomorphism, matter-spirit dualism—are the bankrupt naming schemes of an earlier era.

Today we have a scientific story of creation, but we have not yet learned how to connect the new story to our search for spiritual fulfillment. The cultural historian Thomas Berry looks for the roots of our difficulty in the Black Death of the late Middle Ages. The plague that began in Constantinople in 1334 killed off one-third to one-half of the population of Europe within twenty years. Subsequent visitations likewise decimated the population. It is difficult for us to grasp the catastrophic dimensions of the dying. It was certainly one of the most traumatic events in human history. There were two responses to the trauma, says Berry: One part of the community sought to enlist the intervention of supernatural forces in a world that seemed increasingly hostile to humans; another part of the community sought to remedy earthly terror by understanding earthly process. The first response led to forms of religious faith that emphasize the world's need for redemption; the second led to science. The first sees the material world as the enemy; the second sees matter as an amoral matrix upon which we can impose our will. The first looks for solace in prayer, miracles, and magic; the second in technique. We have not yet resolved this split in our culture, says Berry. We have a new scientific creation story, but we have not integrated the story into our lives as believers and seekers. No community can exist, he says, without a community story.[1]

Our society is divided into two segments. One segment rejects the scientific creation story in favor of empirically obsolete stories that at least make us feel good. The other segment embraces the scientific story but languishes without a collective means for satisfying deep-seated emotional needs or for celebrating the creation. The antagonisms between the two segments of society are deeper than they appear to be, says Berry: "This is precisely why communication between these two is so unsatisfying. No sustaining values have emerged. The problems of the human are not resolved. The human adventure is not dynamized."

Abacus and Rose Combined

If the new story of creation provided by science is not what we want to hear, then what are we to do? Admittedly, it is not easy for a creature who has long thought of himself as the central and immortal apex of creation—the apple of God's eye—to accept that he is unexceptional, contingent, and ephemeral in the cosmological scheme of things. But if this last is true—and everything we have learned about the world suggests that it is—then in turning away from the truth we turn away from the creation and therefore from the Creator, exalting ourselves and our hopes *of what might be* into a false idol.

"Put on your jumping shoes," cried the fourteenth-century mystic Meister Eckhart, "which are intellect and love." Religion without science is idolatrous. Science without religion might be even more dangerous: amoral power without constraint, without wisdom, without love. What is required—if not in each of us individually, then in our collective expressions of faith—is a unity of knowing and believing described in a sonnet of the mathematician Jacob Bronowski:

I, having built a house, reject
The feud of eye and intellect
And find in my experience proof
One pleasure runs from root to roof.
One thrust along a streamline arches
The sudden star, the budding larches.

The force that makes the winter grow
Its feathered hexagons of snow,
And drives the bee to match at home
Their calculated honeycomb,
Is abacus and rose combined.
An icy sweetness fills my mind,

A sense that under thing and wing
Lies, taut but living, coiled the spring.[2]

The venerable Talmudist Joseph Soloveitchik describes two human types, which he calls Adam I and Adam II, corresponding to the Adam of the first and second chapters of Genesis, respectively. Adam I is driven by curiosity. He wants to know how the cosmos works. His command is to "fill the earth and subdue it." His contemporary representatives are the scientist, mathematician, technologist, and secular intellectual. Adam II is also bewitched by the cosmos, says Soloveitchik, but "looks for the image of God . . . in every beam of light, in every bud and blossom, in the morning breeze and the stillness of a starlit evening." His contemporary representative is the mystic, the poet, the ascetic, the person of faith. Adam I is secular by nature, uninterested in questions that cannot be answered empirically; Adam II is more introspective, more spiritual, trusting his intuition of the divine. Adam I seeks mastery over nature; Adam II wishes to be overpowered by nature. Although Soloveitchik clearly identifies himself with Adam II, he knows that Adam I also follows God's command. Like Bronowski, Soloveitchik rejects the feud of eye and intellect, and knows that the completion of creation requires the energies of both Adams.[3]

One Vision, One Knowledge, One Love

"The eye by which I see God is the same eye by which he sees me," said Meister Eckhart. "My eye and the eye of God are one eye, one vision, one knowledge, one love." Like Hildegard of Bingen, Francis of Assisi, Mechtild of Magdeburg, Julian of Norwich, Nicholas of Cusa, Gerard Manley Hopkins, and many others, Meister Eckhart was a mystic-poet of the creation. Shortly after his death in 1327, he was condemned by the Inquisition as a pantheist. The condemnation appears to have been politically motivated; certainly his judges had not read all of his work. According to C. F. Kelley, it is now generally agreed by most Eckhartian scholars "that had the fourteenth-century authorities intelligently and dispassionately investigated all the Meister's writings, he would probably never have suffered condemnation."[4]

However, the Inquisition was not far off the mark. For Eckhart, every animal, every plant, every stone is a revelation of the divine. He was not a pantheist, but neither was he quite exactly an orthodox theist; that is, he did not identify God with the visible world, but neither did he imagine a God who exists outside of creation. Eckhart's God and the creation are separate but inseparable—a faith that (following theologian Matthew Fox) might be called pan*en*theistic: all things in God, God in all things. This is a subtle distinction, the stuff of Inquisitional nitpicking, and I don't want to make too much of it, but it characterizes a long tradition of creation spirituality in Western thought—a tradition that has often been suppressed in favor of redemption theology. Especially since about the time of Eckhart's condemnation, the dominant Judeo-Christian tradition has understood the material world to be intrinsically evil, stained by sin, in need of salvation. Such a tradition is unlikely to look upon the creation as revealing the divine face.

To the extent that God exists outside of creation, he is unknowable. If we are to find him, we need not go into the desert or to the mountaintop—or to gurus, prophets, shamans, or dreams. He is everywhere revealed in all that exists, and nowhere more powerfully, it seems to me, than in the molecular machinery of life and consciousness. The God of my early religious training—the gray-bearded version of ourselves that we see looking down from the ceiling of the Sistine Chapel—turned sticks into serpents, water into wine. As we have seen in an earlier chapter, the God of DNA turns single microscopic cells into untutored shorebirds that find their way unerringly from northern Canada to Tierra del Fuego. The flight of the red knot is no one-time miracle. It is a continuing miracle, and the more of it we understand, the more miraculous it seems.

The God of my early religious training pulled off tricks that are not beyond the powers of any competent conjurer; Harry Houdini or David Copperfield could turn a stick into a serpent or water into wine without batting an eye. But no Houdini or Copperfield can turn microscopic cells into a flock of birds and send them flying on their planet-spanning course. No Houdini or Copperfield can cause consciousness to flare out and embrace the eons and the galaxies. The dubious miracles of the scriptures and of the saints are an uncertain basis upon which to base a faith; the greater miracle of creation is with us twenty-four hours a day, revealed by science on every side, deepening and consolidating our sense of awe.

"God is *here*—in this very place—just as much incarnate as in a human being long ago," said Meister Eckhart. If we want to know who we are, where we came from, and why we are here—the big questions, the cosmological questions—we must turn to the creation as we understand it today, rather than to 10,000-year-old myths. When we are seriously ill, most of us will consult a modern medical doctor rather than a folk healer who offers leeches and herbal concoctions, confident that the new knowledge is more reliable than the old. But when it comes to religion, many of us stick to archaic cosmologies that would evoke condescending smiles if they were the basis of faith for anyone but ourselves.

What does contemporary science say about the big questions? *Who am I? Where did I come from? Why am I here?* The answers at first blush are not what we want to hear: *We are staggeringly complex electrochemical machines. There is no ghost in the machine, no soul that exists independently of the body, and therefore no self that will survive the body's disintegration. Our bodies, minds, and consciousness evolved over hundreds of millions of years from primitive organisms, on a planet that formed from a gassy nebula about 4.5 billion years ago near a star that is just one of a trillion stars in the Milky Way Galaxy. The Milky Way is typical of tens of billions of visible galaxies that probably had their beginning 10 to 15 billion years ago in a cataclysmic explosion from a seed of infinite energy. There appears to be nothing central or special about who, what, or where we are in the universe of galaxies. We are contingent throw-offs of organic evolution, at least in the details.* And why are we here? *We are here to make copies of our genes and thereby ensure the continuance of our species after the deaths of our selves.*

Is there more to it than this? Of course. We love, we laugh, we fight, we cry. We value good and avoid evil. We wish happiness for our children and for our children's children. We are creators of art and poetry and forms of religious worship. We hold the fate of the planet in our hands. And this too: *In the human self the universe becomes conscious of itself.*

Our intellects embrace the DNA and the galaxies. Our scientific knowledge of the world is tentative, partial, evolving; it is a human creation, flawed with the limitations of our race, hedged about by our paltry powers of perception and cognition. Nevertheless, it is the most reliable knowledge of reality that we have. If anyone doubts that scientific knowledge is reliable, I will take the DNA of the red knot, adjust the four-letter code, and send the bird winging on a new course. (Molecular biologists perform similar feats every day.) If anyone doubts that our telescopes reliably reveal the galaxies in their august places, let them follow the journeys of the Voyager spacecrafts out beyond the farthest planets on trajectories precisely calculated in advance. The reliability of scientific knowledge is confirmed every day, all around us, in the accoutrements of technological civilization.

Science is knowledge that has had its feet held to the fire. It is knowledge we can count on. We can make of ourselves whatever we want, we can use our knowledge for good or ill, but our future will be best assured if what we choose to make of ourselves is consistent with what we know to be reliably true.

EIGHT

Creationism and Evolution

Nothing makes sense except in the light of evolution.

Theodosius Dobzhansky

The alternative to thinking in evolutionary terms is not to think at all.

Sir Peter Medewar

THERE ARE PEOPLE who prefer that things remain the same; there are people who like change. We call them respectively reactionaries and progressives. Charles Darwin was a reluctant progressive. In early 1844, eight years after the voyage of the *Beagle*, he wrote to his friend Joseph Hooker, "I am convinced (quite contrary to the opinion I started with) that species are not (it is like confessing a murder) immutable."

Species change! Most species that existed in the past have become extinct. New species will exist in the future, having descended from species that exist today. All present animals and plants (including ourselves) are but a momentary snapshot in an ever-changing flux of life. Here was one of the most progressive ideas in the history of human thought, and Darwin was reluctant to admit it. To think such a thing was akin to murder. He would delay publication of his great book for sixteen years, then publish only when his priority was threatened by Alfred Wallace.

Darwin was keenly aware of the political, social, and religious implications of his new idea. If species change, then so might established institutions: the church, the landed gentry, the ruling class. Religion, especially, appeared to have much to lose, grounded as it was in scriptural cosmology. If the Bible was wrong in the very first chapter of Genesis, then the veracity of the entire enterprise was called into question. Evolution was not just a scientific idea, it was a bombshell, taken up by progressives as justification for social reform, damned by reactionaries as subversive of the social order, welcomed by atheists, feared by theists. It pitted against each other the two great hankerings of humankind—for fixity and for change—and the corresponding postures of mind—True Believer and Skeptic.

Darwin tells us over and over that he resisted the idea of evolution. The mutability of species was forced upon him, he says, by the evidence of nature. If evolution by natural selection has revolutionary consequences, it was nature's fault, not his. The lesson of change was whispered in Darwin's ear when he visited the Galápagos Islands in 1835 aboard HMS *Beagle*. These are young volcanic islands, which emerged from the sea within the last few million years. They contain a variety of closely related species of plants and animals, similar to those on the South American mainland 600 miles away, yet different—species found nowhere else on Earth and perfectly adapted for life in the islands. The patterns of evolution by natural selection were there to be seen, in stark simplicity, uncluttered by the comings and goings, stirrings and blurrings of the mainland. But the lesson of the Galápagos had to germinate in Darwin's mind, slowly overwhelming his innate conservatism, until finally the truth became irresistible: Immigrant plants and animals from the mainland had adapted to a new environment, becoming new species in the process.

Darwin was not an ardent Skeptic, but neither was he a True Believer. Evolution was *forced* upon him by his meticulous examination of the evidence.

Everything in his life—his social position, his religiously conservative wife, his personal temperament—inclined him toward established doctrine. As Francis Bacon said, what we would like to be true, we preferentially believe, but Darwin, at least, was willing to let nature have its say. He chose truth rather than peace of mind.

Now, more than a century later, we are still locked in a battle between those who believe what makes them feel good and those who trust the evidence of their senses. The battle lines are drawn in about the same place as they were in the immediate aftermath of the publication of Darwin's *Origin of Species*. At issue are two great worldviews: In the one, humans are coeval with the world itself and reside at its center; in the other, the human species is a recent and contingent offshoot of evolution in a typical corner of a universe that is vast beyond our knowing. As we have seen, Americans are divided down the middle on this issue. Evolution of species by natural selection remains the great bugbear of religious fundamentalists, the issue upon which they have chosen to stake all. If the nonfixity of species is admitted, then their whole edifice of True Belief comes tumbling down.

Biologists vigorously debate the details of evolution and the sufficiency of natural selection as the agent of change, but no one within the scientific community doubts that life evolved over billions of years from simple beginnings, that all life is related by common descent, and that humans are part of the web of life. Most scientists would say that these statements are facts, not theories—or at least as close to being facts as our rational instruments of knowing can make them. The two epigraphs that begin this chapter, the first by one of our greatest geneticists, the second by a Nobel laureate biologist, give a sense of evolution's gold-standard currency within the scientific community. In the fall of 1996, even the conservative Roman Catholic pope John Paul II asserted that evolution is "more than a theory."[1] In fact, the issue is so cut-and-dried I feel a little embarrassed even talking about it. However, not to talk about it is to put science teaching in our public schools at risk. The teaching of evolutionary biology is under nationwide assault by fundamentalist Christians, led by the powerful Traditional Values Coalition, a group that represents thousands of conservative churches.

Warm and Fuzzy Science?

Reverend Louis Sheldon, spokesperson for the coalition, was quoted in the press as saying: "When they teach kids that they came from monkeys, that's a dead dinosaurial kind of thing. It's a negative. It's not a warm, fuzzy kind of thing."[2] This curious remark brings to mind some suggestive experiments done with baby monkeys in the 1950s by the American psychologist Harry Harlow.[3]

Harlow wanted to discover if mother love in infants is generated by the satisfactions of feeding, as had long been postulated by psychologists. He placed eight infant monkeys in individual cages, each with equal access to two surrogate "mothers." One mother was made from bare welded wire surmounted by a wooden head. The other mother was comfortably sheathed with soft terry cloth. Four of the infants received their nourishment from their wire mother, and four from their cloth mother, the milk being furnished in each case by a nursing bottle with its nipple protruding from the mother's "breast."

The wire and cloth mothers proved to be *physiologically* equivalent: All of the infants drank the same amount of milk and gained the same weight. But the mothers were not *psychologically* equivalent. The infants that took their nourishment from a wire mother spent no more time with her than feeding required. All of the infants spent most of their nonfeeding time clinging and cuddling to their warm and fuzzy cloth mothers. It is apparently not satisfactions of feeding that generate affection in infants, but close bodily contact and cozy security.

Harlow then exposed the infant monkeys to a room that was far larger than the cages to which they were accustomed. He placed in the room a number of unfamiliar objects, such as a small artificial tree, a crumpled piece of paper, a folded gauze diaper, a wooden block, a doorknob. He reports: "If the cloth mother was in the room, the infant would rush wildly to her, climb upon her, rub against her and cling to her tightly. Its fear then sharply diminished or vanished. . . . If the cloth mother was absent, however, the infants would rush across the test room and throw themselves face-down on the floor, clutching their heads and bodies and screaming their distress. . . . The bare wire mother provided no more reassurance in this 'open field test' than no mother at all."

The *metaphorical* possibilities for interpreting these experiments are irresistible, especially in light of the remark by the spokesperson for the Traditional Values Coalition. It is tempting to see the wire mother with the milk-producing nipple as evolutionary science, and the terry-cloth mother as fundamentalist religion. In the late twentieth century, science is the source of our health, wealth, and physical well-being but provides little in the way of emotional comfort. As a culture, we divide our time between science and religion, going to the former when in need of physical sustenance (technology, medicine, creation of wealth), but spending most of our time clinging to the latter. When faced with a large and frightening universe filled with unfamiliar objects, we do not turn to science for reassurance, but to the "warm and fuzzy" truths of fundamentalist faith.

Can I take the metaphor further? Growing up has something to do with putting aside our teddy bears and security blankets. The spokesperson for the Traditional Values Coalition underestimates our children when he insists that schoolkids can't handle cold and clammy truths like descent from reptilian or amoebic ancestors. It would be comforting to think, as did our ancestors, that we live in a nurturing universe, centered upon ourselves, embraced by nearby, consoling stars. The truth, however, is rather different. Our Earth is a typical planet in an "infinite immensity of spaces," and it is a measure of our adulthood that we have the courage to accept this more difficult truth. It is comforting to imagine that our species had its origin in a secure nursery—a garden perhaps—presided over by a watchful parent, but again the truth turns out to be otherwise.

Evolution is not warm and fuzzy. It can even be capricious and sometimes cruel. It does, however, have much that recommends it to the adult mind; it is a *fact* by every criterion of science. Our schoolkids do not need intellectual security blankets. By insisting that science textbooks be warm and fuzzy, fundamentalists encourage the infantization of the next generation of Americans.

Facts and Theories

I grew up in the Bible Belt, not far from Dayton, Tennessee, the site of the Scopes monkey trial. Early on I moved to New England so that my children could be raised in the thoughtful tradition of the Adamses, Emerson, Thoreau, Agassiz, and Gray. (I was influenced in this decision by two books in my mother's library when I was a child: Van Wyck Brooks's *The Flowering of New England* and *New England: Indian Summer*.) Now the Bible Belt has been loosened to encompass the expanding girth of fundamentalism. Even in New England, strident voices on school committees are agitating to have "creation science" taught in the classroom along with the "theory" of evolution.

There are some fundamental misapprehensions at work here. First, there is the so-called opposition of "fact" and "theory." Scientists speak of any group of related assertions about the world as a "theory." Some theories are firmly held, such as the atomic theory of matter or the cell theory of life, so firmly held that atoms and cells are spoken of as facts. Other theories are highly speculative, such as the theory that quasars are black holes forming at the centers of galaxies in the early universe; astronomers are cautious about calling galactic black holes "facts." There is much about the theory of evolution that scientists would call 99.9 percent fact, such as the idea that life developed on this planet over hundreds of millions of years from simple beginnings. There is much about the theory of evolution that remains speculative, including the pace of change and the sufficiency of natural selection as a driving force. These topics are hotly debated by evolutionary biologists.

As noted earlier, the idea that life was created essentially as it is today some time within the last 10,000 years has zero standing as scientific fact. I know of no research anywhere in the peer-reviewed scientific literature that supports such a theory. Of course, creationists claim that science is ruled by dogma, that dissenting views are not given a fair hearing. Nothing could be further from the truth. If there were any solid, reproducible evidence supporting a young Earth, scientists would be falling over each other to publish it. Being on top of a revolutionary and successful idea is the way scientific careers are made. Every scientist I know is as happy to have something proved wrong as proved right. Either outcome advances us toward truth.

An instructive example of how science *actually* works is afforded by an article called "Phylogeny Reconstruction and the Tempo of Speciation in Cheilostome Bryozoa," published in 1994, by Jeremy

Jackson and Alan Cheetham, both associated with the Smithsonian Institution.[4]

First, a little background. Darwin believed that species arise by the gradual and *continuous* accumulation of tiny changes within isolated populations of animals or plants; for example, mainland invaders of the newly formed Galápagos Islands slowly diverge from their ancestral populations on the mainland until they are different enough so that they can no longer interbreed.

In 1972, evolutionary biologists Niles Eldredge and Stephen Gould introduced a new idea, called "punctuated equilibrium." They proposed that new species appear rapidly in the geological record, presumably at times of environmental change or stress, followed by long periods when species remain relatively unchanged; continuing with the same example, while the Galápagos invaders rapidly evolved to adapt to their new environments, their well-adapted stay-at-home mainland cousins remained relatively unchanged.

Alan Cheetham, coauthor of the 1994 article, had long been opposed to punctuated equilibrium, championing instead the classic Darwinian idea that populations evolve gradually and continuously through millions of years of natural selection, occasionally giving rise to new species. With his colleague Jackson, he set out to provide the most detailed reading yet of the fossil record of Bryozoa (coral-like animals well represented in the fossil record), confident that the Darwinian view would be sustained. He writes: "I came reluctantly to the conclusion that I wasn't finding evidence for gradualism." What he found instead were individual species of Bryozoa persisting virtually unchanged for millions of years, then in brief moments of geologic time giving rise to new species. Cheetham switched his allegiance to punctuated equilibrium. The lesson is clear: *What we want to believe is not necessarily true.* The only way to discern the truth is to observe nature. Like Darwin, Cheetham was won over by the evidence of nature.

Personal Incredulity

Let's pursue these issues a bit further. The human eye is dear to creationists. It is so exquisitely suited for its purpose—to provide sharp visual images of the outside world to the brain—that no sequence of random variations acting over time would seem sufficient for its design. Michael Pitman, an influential creationist writer, says, "That such an instrument should undergo a succession of blind but lucky accidents which by necessity led to perfect sight is as credible as if all the letters of the Origin of Species, being placed in a box, shaken and poured out, should at last come together in the order in which they occur in that diverting work." This is typical of what creationists want to teach our children in public school science classes. It's an argument I often hear from respondents to my occasional *Globe* columns on evolution. Like other creationists, Pitman wonders how natural selection could have favored anything less than the perfect organ. "The eye must be perfect or near perfect," he writes. "Otherwise, it is useless."[5]

My own eyes are far from perfect; without glasses I could not read Pitman's diverting book. Nevertheless, my imperfect eyes are not useless. Even the ability to vaguely perceive light and dark makes life easier. The one-celled marine organism *Euglena* has an eyespot containing a few specks of pigment by which it orients itself toward the light, which it uses to manufacture nutrients by photosynthesis. *Euglena*'s eyespot is hardly a perfect visual instrument by human standards, yet it is crucial to the organism's survival. The scallop has dozens of eyespots, each with its own lens and light-sensitive cells. These organs are crude, but they serve the scallop well as it scoots across the ocean floor. A scallop without eyespots would be at a considerable disadvantage in the struggle for life. The variety of eyes in nature is astonishing. Virtually every image-forming method devised by human technology has been anticipated by nature: lenses, mirrors, pinhole cameras, and fiber-optic bundles. Eyes of one sort or another have independently evolved at least forty times during the history of life.

Of course, creationists are most enamored with the eyes of the higher vertebrates, including ourselves. The human eye, with its lid, lashes, adjustable iris, adjustable lens, and 125 million light-sensitive receptors, is considered by creationists to be the crowning glory of the creation. A creationist writer, I. L. Cohen, says of the human eye: "The whole system came into being at precisely the same time. . . . It is not possible, under any stretch of the imagination, that all these parts of the optical mechanisms could haphazardly become functional through random mutations."[6]

Not possible? Under any stretch of the imagination? This is the sort of thinking that biologist Richard Dawkins calls "The Argument from Personal Incredulity": *If it seems impossible to me, then it must be impossible.* This hugely anthropocentric fallacy assumes that nature conforms to the limitations of our imaginations. The Argument from Personal Incredulity is a kind of idolatry: making the Creator in our own image. The only way to decide whether something is possible or impossible (I am not talking about logical contradictions) is to go to nature and observe. Or, as is increasingly the case in science today, to model natural processes on a computer.

A computer study by evolutionists Dan Nilsson

and Susanne Pelger suggests that the evolution of the eye by natural selection may be less difficult than previously estimated even by biologists.[7] They modeled evolution of the eye computationally. They started with something akin to an eyespot, a flat patch of light-sensitive cells sandwiched between a transparent protective layer and a layer of dark pigment, equivalent to the simple eyespot of *Euglena*. They allowed the eyespot to deform itself at random, with the requirement that any change be only 1 percent bigger or smaller than what went before. They also provided for random changes in refractive index of the transparent layer. (Refractive index is a measure of the speed of light in a medium compared to the speed in a vacuum, and affects the amount an oblique ray of light is bent upon entering the medium.) The image quality at each step was calculated using elementary optics. The two researchers made assumptions about heritability and intensity of natural selection based on research with living species in the field, choosing the most conservative numbers in each case. They then set the computer program running and observed the results.

Here is Richard Dawkins's description of what happened: "The results were swift and decisive. A trajectory of steadily improving acuity led unhesitatingly from the flat beginning through a shallow cup to a steady deepening cup. The transparent layer thickened to fill the cup and smoothly curved its outer surface. And then, almost like a conjuring trick, a portion of the transparent filling condensed into a local, spherical subregion of higher refractive index."[8] In other words, an eye socket, a curved retina, and a lens appeared on the screen of the computer. Using the most conservative assumptions about how changes are propagated through offspring, the researchers found that the time taken to evolve a vertebrate eye from a flat patch of light-sensitive skin was 400,000 generations. That's half a million years or so for typical small animals, a mere blink of the eye in geological time.

Charles Darwin himself expressed occasional doubts that natural selection could have produced the vertebrate eye, even with millions of years to work with. However, as Dawkins suggests, personal incredulity is not a reliable guide to truth. What seemed unlikely to Darwin, and seems impossible to creationists, has been shown to be quite reasonable by high-speed computer modeling. Not only reasonable, but given the proven premises of random mutations and natural selection, virtually inevitable. Will successful computer simulations make any difference to creationist True Believers? Not likely. Although creationists claim they want to expose our schoolchildren to alternate "scientific" accounts of creation, there is nothing remotely scientific about the Argument from Personal Incredulity. All we learn from the argument is that the Creator is more resourceful than the creature can possibly imagine.

God of the Gaps

Charles Darwin was in the Galápagos Islands only for a matter of days. He did not observe the flora and fauna of the islands evolve. His theory of evolution by natural selection was based on logical deduction, not observation. In the mid-1800s, no one, not even Darwin, had ever seen a new species emerge in nature. Evolution, as Darwin imagined it, takes place on a time scale that is vast compared to human experience. For that reason, it has long been thought impossible to observe natural selection in action. Creationists make much of this fact to emphasize that evolution is "just a theory."

By the same argument, it is "just a theory" that there once lived a person named George Washington who was the first president of the United States. Evolution is a historical science, confirmed by historical evidence, primarily the fossil record. For a long time, even biologists were inclined to admit that evolution might *never* be an experimental science. Now it turns out that evolution is happening all around us, day by day—most dramatically and dangerously in the case of pathogenic bacteria evolving resistance to antibiotics. For example, the bacterial agent of malaria has evolved resistance to many of the drugs that formerly held that disease in check. This is compounded by the evolution of DDT resistance by the mosquito that carries the malaria pathogen during part of its life cycle. After a period during which malaria worldwide was on the decline, the disease is now making a roaring comeback as the world's biggest killer of children. Creationists who deny evolution not only contribute nothing to the resolution of the malaria problem but also undermine the scientific education that will help the next generation solve the problem.

Creationists who doubt the reality of evolution should read Jonathan Weiner's wonderful book *The Beak of the Finch: A Story of Evolution in Our Time.*[9] The central protagonists are Peter and Rosemary Grant, of Princeton University, who for twenty years have observed generation after generation of finches on one of Darwin's Galápagos Islands—measuring, weighing, tracking, and analyzing the birds' struggle for existence. The Grants have come to know every one of thousands of birds individually. They have watched populations evolve—yes, physically evolve—in times of stress and in times of plenty. It is a brilliant story, beautifully told, which stunningly confirms in scope and particulars Charles Darwin's reluctantly progressive leap of faith. Weiner also catalogs ongoing evolution observed by scientists in other places: guppies in the Caribbean, soapberry bugs in the American South, stickleback fish in ponds of the Canadian West, and, of course, bacteria worldwide.

Evolution has at last become an experimental science. It cannot be long before we will observe the holy grail of experimental evolutionary studies, the emergence of a new vertebrate species by natural selection under controlled conditions.

The Institute for Creation Research (ICR) in California is this country's most prolific font of religion-inspired antievolution propaganda. "These are scientists," one of my *Globe* correspondents insisted, "and they are finding evidence that evolution is wrong." Nonsense. ICR "scientists" are conspicuously deficient in primary research and publish nothing in peer-reviewed scientific journals. They have negligible standing within the scientific community. The real scientists are the ones like Peter and Rosemary Grant or Jeremy Jackson and Alan Cheetham who get out in the field or go into the lab and do the nitty-gritty work that is always involved in asking nature an answerable question. What the ICR crowd is best at is cataloging gaps or supposed inconsistencies in evolutionary science. (The "impossibility" of the eye evolving by natural selection is an example of a supposed inconsistency.) Any evolutionary biologist can list as many gaps in our knowledge as can creationists, and probably more. If there were no gaps, science would be at an end.

Scientists look at the overwhelming success of evolutionary science and assume that the gaps will be filled as our knowledge becomes more complete. Creationists point gleefully to the gaps: "See," they shout, "evolution is a shambles." They assume that because we don't know everything, we know nothing.

Science is a dynamic social activity, made up of millions of men and women of all religious faiths, races, nationalities, and political persuasions. It is preposterous to suggest, as do creationists, that this vast and diverse assemblage of scientists, many of them devoutly religious, is guided by blind commitment to Darwinian dogma. As noted earlier, the evolution of life over hundreds of millions of years has virtually 100 percent support of the organized scientific community, whereas biblical creationism has essentially zero support. To suggest that creationism should get equal billing in our public schools not only is unconstitutional (violating separation of Church and State) but is simply silly. One might as well give equal billing to those who believe the Earth is flat.

By supposing that evolutionary science and religion are in conflict, creationists impoverish religion. Science reveals a creation story worthy of a Creator who is more, much more, than a warm and fuzzy projection of ourselves. If we are to find him, we had best look to the creation rather than into the dusty mirror of Personal Incredulity. If we want knowledge of the creation, our best instructors are the researchers who spend years scrambling across Galápagos rocks tagging finches or in the laboratory patiently measuring and cataloging thousands of fossils, rather than to those nit-picking creationists doing secondhand "science" in California.

Life is a flame that dances on the face of creation, never still, infinitely creative, with a prodigiously creative past and, it seems to me, bright potential for the future. I guess that makes me progressive in temperament. Others of a more reactionary bent will continue to plump for fixity of species, regardless of the progress of empirical research, clinging to their terry-cloth mother of archaic cosmology for all it's worth. Attitudes toward evolution can be as much a matter of temperament as of evidence.

NINE

Academy in Revolt

The only source of knowledge is experience.

Albert Einstein

ELSA EINSTEIN WAS once asked if she understood her famous husband's theory of relativity. She replied, "Oh, no, although he has explained it to me many times—but it is not necessary for my happiness."

For what, then, *is* science necessary? ask a growing number of academic critics of science. For most of its history, science has counted on the support of the intellectual establishment while criticism of science has come mostly from religious fundamentalists and literary romantics. One hundred years ago, Andrew Dickson White wrote a big, two-volume *History of the Warfare Between Science and Theology in Christendom.* He had much to write about, and little has changed since then. Fundamentalist Christians continue to reject "scientific atheism" (or alternatively, "secular humanism") with the old fervor. Even mainline Christian churches still struggle uncomfortably to come to some sort of accommodation with science.

The romantic assault on science was established early in the nineteenth century by Keats (". . . Do not all charms fly / At the mere touch of cold philosophy?") and Wordsworth ("We murder to dissect . . ."), and was amplified by later poets such as Emily Dickinson, Vachel Lindsay, e. e. cummings, and others. The gentle needling of science by poets has never been more than a minor nuisance, and in recent times seems to have abated. Lately, however, old intellectual allies have turned against science, attacking the Enlightenment enterprise from *inside* the academy, in some cases explicitly repudiating the legacy of the Enlightenment and introducing a dangerous new species of ideological True Belief.

Typical of the new criticism is Bryan Appleyard's *Understanding the Present*, a lively and erudite subversion of science by the sort of fellow one might expect to be a champion of the Enlightenment. Appleyard is a special features writer for London's *Sunday Times* and the author of several books on modern culture. He is broadly knowledgeable about the history and philosophy of science, and is a nimble rhetorician. Not only is science unnecessary to our happiness, he argues, it is positively inimical to it. "Science, quietly and inexplicitly, is talking us into abandoning our true selves," he writes.[1]

The editors of *Nature*, Britain's most influential science periodical, called Appleyard's book an "assault upon reason" and pronounced it "dangerous." According to the editors, Appleyard's slick polemic plays into the hands of those right-wing politicians who blame supposedly novel social ills on erosion of traditional religious values by secular scientific humanism. These are often the same politicians who control the purse strings of science's public support. Nothing pleases them more than intellectual benediction upon their know-nothing reactionism.[2]

As an analysis of the spiritual malaise of our times, Appleyard's critique of science is provocative but ultimately unsatisfying. Science is intrinsically domineering, all-pervasive, and totally incompatible with religion, he writes. He admits that science is effective, but asks: What does it tell us about ourselves and how we must live? His answer: Nothing. Appleyard recoils from the conclusion, widely embraced within the scientific community, that we are cosmically inconsequential bundles of stardust, adrift in an infinite and purposeless universe. He affirms our specialness, as an act of faith, and asserts, again as an act of faith, that the true self is forever beyond the reach of scientific explanation (a conclusion I will examine closely in the next chapter).

The trouble began, says Appleyard, with Galileo's defiant claim that the Earth is not the center of the universe, and things have been going downhill ever since. He implies that we would be better off if Galileo, Newton, Darwin, and Freud had kept their mouths shut and subverted facts to the traditional sense of our unique importance in the universe. When facts and meaning are in conflict, Appleyard chooses meaning over fact. There is a self-imposed ignorance about Appleyard's choice that would be merely risible if it were not so well expressed. He makes his choice, he says, with Kierkegaard, "on the basis of ourselves in spite of, even in opposition to, the facts of the world." One can admire the boldness of his choice—it takes a certain perverse courage for a contemporary intellectual to profess faith over fact—while recognizing that it is a choice that we, as a culture, must not make. The alternative to "facts" is not a Golden Age of good feeling and theistic certainty, but a return to the pre-Galilean world of autocracy, vulnerability to disease, sectarian violence, and rank superstition.

Appleyard's chief bugbear is something he calls "scientific liberalism," an attitude of mind that rejects *absolute* belief in anything. What he means, of course, is old-fashioned skepticism. What scientists consider virtues—skepticism, tolerance for the views of others, a belief in progress—Appleyard sees as agents for the erosion of traditional religion and absolute moral values. He is right about religion and values; the relevant modifiers here are *traditional* and *absolute*. But skepticism, tolerance, and belief in progress are certainly *not* in opposition to spirituality and ethical behavior. It is a thesis of this book that these qualities *must* be associated with any religion that will lead us out of our present spiritual malaise and provide a satisfactory moral compass for our inevitably scientific and technological future.

There is nothing particularly new in Appleyard's critique. Similar views are held by the majority of Americans. Scientific skepticism has long been the favorite whipping boy of religious fundamentalists and right-wing ideologues of all stripes. With Appleyard, however, these subversive views come wrapped in a glitzy veneer of establishment sophistication. To put it bluntly, Appleyard's subversion of science is an inside job, and it is probably this more than anything else that rankled the editors of *Nature*. Their knee-jerk reaction to his book was perhaps misplaced. Appleyard is on target in much of what he says about science. Our new understanding of the cosmos does indeed rest uncomfortably with the traditional view that we are the point and purpose of creation. If there is a self that is more than the arrangement and interactions of our atoms, science has not found it—nor does it expect to find it.

As Appleyard correctly apprises, everything science has learned since Galileo suggests that we are accidental, contingent, ephemeral parts of creation, rather than lords over it. Such news can be difficult to accept for creatures who long believed themselves to be the raison d'être of the universe, and Appleyard is one of those creatures. He would have us roll back history to the epochal confrontation of Galileo and the Church, throwing in our weight with the ecclesiastical authorities who argued that scientific theories are mere useful fictions, of lesser reliability than the truths of revelation and Church tradition. To suppose that the Earth revolves about the Sun may be a handy way to calculate the calendar, but we shouldn't confuse heliocentric astronomy with the truth, said Galileo's persecutors; if a scientific idea contributes to our economic or physical well-being, then by all means use it, but don't forget that it is only a clever means to an end, with no more claim on our spiritual lives than a water mill or a mechanical clock.

Of course, Galileo saw it differently. Following his forced denial of the Earth's motion—as an old man on his knees before assembled cardinals—legend claims he murmured under his breath, "Yet it moves." Scientists, of course, will agree with the whispered remark. More to the point, celestial physics has amply confirmed to all but the invincibly ignorant that heliocentrism is *fact*.

Most scientists also reject the link, asserted by Appleyard, between the ascendancy of science and the spiritual malaise of our time. Our diminished sense of the sacred has resulted not from the growth of knowledge but from the failure of traditional religions to incorporate scientific discovery into a framework of spirituality and religious worship. Appleyard is right to choose the Galileo case as a defining moment in our history. At the time of Galileo's trial, the Church could have gone either way: It could have embraced the new knowledge of the cosmos and infused it with sacred meaning, or it could have kept its wagon firmly hitched to the weary nag of Aristotelian physics. We know which choice was made. The Church opted for a defunct cosmology—animistic, supernaturalistic, geocentric. It failed to grasp Galileo's skeptical lesson that *no* human knowledge can capture the essence of the divine, a lesson reiterated by Einstein 300 years later: "Science not only purifies the religious impulse of the dross of its anthropomorphism but also contributes to a religious spiritualization of our understanding of life." No sentiment more opposite to Appleyard's could be found. For Galileo and Einstein, fact and meaning *cannot* be in conflict.

Bryan Appleyard's vexatious misunderstanding of the present is awash with unwarranted pessimism and nostalgia for an Eden of anthropocentric belief. His prescribed cure for our spiritual malaise is a blinkered rejection of the very things that make us human: curiosity and reason. In a world beset by sectarianism and ethnic and racial intolerance, a little "scientific liberalism" can serve us well. Understanding relativity may not be necessary to our happiness, but an open, tolerant, joyful search for truth most certainly is.

The Unalterable Consistency of Fact and Meaning

A yet more devastating critique of contemporary science was delivered by Václav Havel, a Czech poet, playwright, and statesman, as an address before the World Economic Forum titled "The End of the Modern Era." Havel singles out "rational, cognitive thinking" and "depersonalized objectivity" as the abiding sins of our century. He directs our attention to the belief, which had its beginning in the Renaissance and its culmination in the Enlightenment, that man is capable of "describing, explaining and controlling everything that exists, and of possessing the one and only truth about the world." This, he asserts, is the bankrupt agenda of modern science. Science can describe the different ways we might destroy ourselves, he says, ominously, but it cannot offer us effective and practical instructions on how to avert our destruction. What is required, he suggests, is not some sort of compromise between contending views but a decisive end to the Enlightenment program. We must recognize that the flawed notion of objective, transpersonal truth about the natural world leads to human degradation and the death of God.[3]

This is a shattering accusation—and totally gratuitous. It is certainly true that the post-Enlightenment agenda has been a search for objective, transpersonal truth about the natural world. It is an article of faith of all scientists since Galileo that either the Earth goes around the Sun or it does not, matter is made of atoms or it is not, our bodies are made of cells or they are not, disease is caused by bacteria and viruses or it is not, the human soul survives the death of the body or it does not, and so on. These questions are not matters for a democratic vote, nor are they for politicians to decide. We have an effective, rational instrument for discerning (tentatively, partially, but always more accurately) the facts of the world. Those facts will not always be what we want to hear—what makes us "feel good."

We have learned, for example, that we are dust motes in a cosmos that is vast beyond our knowing. Further, it would appear that we are recent buds on the tree of life, contingent, ephemeral. Our souls are embedded in matter and impermanent. God, if he exists at all, is deaf to our pleas. Are these uncomfortable truths? Yes. Can we live with them wisely and well, retaining a sense of self-esteem? Yes. Can we remain alive to the sacredness of nature and the specialness of human creation? Yes. Can we collectively worship an Absolute that exists beyond all possibility of human knowing, that creates and animates the universe? Yes. Can we live morally, with a shared sense of responsibility to each other and to the planet? Yes. But only if we begin with what reason tells us to be true and redeem our knowledge by our wisdom and our art, accepting the unalterable consistency of fact and meaning.

In this task, poets such as Václav Havel and writers such as Bryan Appleyard should be leading the way, transmuting the dross of scientific fact into the spun gold of sacred meaning. Instead, they ask us to return to the irrational and the subjective, with all that has meant in the way of superstitious fear and factional strife. Havel and Appleyard give voice to a widespread longing for an idyllic past in which gods and humans mingled blissfully in the cheery groves of Acadia, a past which never existed and to which we could not return even if it did.

Suckled in a Creed Outworn

In his book *The Greeks and the Irrational*, E. R. Dodds describes the Greek enlightenment of the sixth century B.C. as characterized by a "progressive replacement of the mythological by rational thinking." By the end of the age of Pericles, however, a fear of the new freedom set in, says Dodds. Supernaturalism returned. Astrology and magical healing replaced astronomy and medicine. Cults flourished, rationalists were scapegoated, and culture began to decline. Dodds calls it "the return of the irrational." He writes: "As the intellectuals withdrew further into a world of their own, the popular mind was left increasingly defenseless . . . and, left without guidance, a growing number relapsed with a sigh of relief into the pleasures and comforts of the primitive. . . . [B]etter the rigid determinism of the astrological Fate than the terrifying burden of daily responsibility."[4]

The Greek experience, described by Dodds, remarkably parallels our own situation. In a 1996 *Time* magazine essay called "The Return of the Primitive," Charles Krauthammer draws attention to the rampant irrationalism of our own time. The world of scientific fact is too much for us, writes Krauthammer. We yearn with Wordsworth "to be a pagan, suckled in a creed outworn." Overwhelmed by the impenetrable cabala of high sci-tech, we turn instead to penny mysteries, to potions, auras, and alternative healing, and give rapt attention to bearers of tales of alien abduction and Satanism in our schools. Krauthammer quotes Oscar Wilde to describe the prevalent view: "I can stand brute force, but brute reason is quite unbearable."

At precisely the moment when we require reasoning intellectuals—theologians, philosophers, social critics, artists—to give us the courage and the critical means to cope with our new knowledge and new independence, these are often the very people who are leading the retreat into primitivism and superstition. In the United States, the academic intellectual left, traditionally allies of rational science against fundamentalism and autocratic authority, have given themselves over to something called "science studies," which asserts (in a common manifestation) that all knowledge is a social construct, either loosely constrained by nature or not constrained at all, and therefore arbitrary, and must be judged by the way in which it serves or does not serve a political agenda—socialist, feminist, environmentalist. As a way of organizing our experience of reality, science is judged a failure, a tool of the repressive establishment, paternalistic, phallic, Eurocentric. At the same time (and inconsistently), these critics ignorantly quote the Heisenberg Uncertainty Principle and the new science of chaos to buttress their absurd contention that in the name of "fact" anything goes.

Many of us who came of scientific age during the 1950s and 1960s were deeply influenced by positivist philosophers such as Rudolf Carnap, Philipp Frank,

Carl Gustav Hempel, Susanne Langer, Robert Lindsay, Ernest Nagel, F. S. C. Northrop, W. V. Quine, Hans Reichenbach, and B. F. Skinner. We were weary of the seemingly endless squabbles of metaphysicians and dreamed of objectivity, even if it meant focusing our attention on the small part of human experience that is amenable to logical analysis. We sought clarity at the cost of completeness. However, the study of the history of science in recent decades has made it abundantly clear that our dream of objectivity was an illusion. Even such confirmed positivists as Albert Einstein created theories that embody personal, institutional, and cultural influences. Certainly science embodies all of the foibles of humankind, and no one inside or outside of science will suppose that the scientific community is without need of improvement. In all of this, the critics of science are right.

But no one who *does* science and *understands* science can doubt that it provides reliable knowledge of the natural world. And this, it seems to me, is what is behind much of the wide popular distrust of science: ignorance. As a people, we live in a civilization that is based upon science, but we know very little about science. A small minority of Americans can be counted scientifically literate. (In this, the scientific community has failed its responsibility to the society that supports it. Scientists must become as forcefully engaged in general education, even at the primary and secondary levels, as they have been engaged in professional education.) Science requires enlightened community criticism, community guardianship, and community engagement, and these must be led by scholars and intellectuals.

What is most distressing about the new academic critics of science is their abysmal ignorance of science, exposed by Paul Gross and Norman Levitt, biologist and mathematician respectively, in their provocative book *Higher Superstition.* Gross and Levitt take on the worst excesses of what they call "the relativism of the social constructivists, the sophomoric skepticism of the postmodernists, the incipient Lysenkoism of the feminist critics, the millennialism of the radical environmentalists, the racial chauvinism of the Afrocentrists," all of whom they lump into the category "academic left," purveyors of "a form of coarse populism that is willing to exile the most stringent kinds of analytical thought and jettison the reliable devices of empiricism in the name of opening the doors of knowledge and driving the haughty priests of science from the temple." Gross and Levitt welcome, even encourage, informed criticism of science, and stress the importance of democratizing scientific and technological decision making; the proper basis for both activities, they insist, is a populace that is well-informed about science, both as a body of knowledge about the world and as a way of knowing. The thrust of their book is to show that much criticism from the academic left is grounded in a profoundly invincible ignorance of the thing criticized.[5]

This ignorance was brilliantly exposed by physicist and social leftist Alan Sokal with a parody of "science studies," submitted to the prestigious leftist journal *Social Text*. Sokal had read Gross and Levitt's *Higher Superstition* and wondered if things were really as bad as the authors claimed. He asked himself, Would a leading journal of cultural studies "publish an article liberally salted with nonsense if (a) it sounded good and (b) it flattered the editor's ideological preconceptions"? Sokal's article, titled "Transgressing the Boundaries: Toward a Transformative Hermeneutics of Quantum Gravity," professed to show that physics has undermined the notion that there exists an "external world, whose properties are independent of any individual human being."[6] This, of course, was just the kind of supposedly scientific justification for their antiscience bias that the editors wanted to hear.

Unfortunately for them, Sokal's article was chockablock with *such* outrageous nonsense that anyone who knew even a little science could only laugh. (An example: "In this way the infinite-dimensional invariance group erodes the distinction between the observer and observed; the π of Euclid and the G of Newton, formerly thought to be constant and universal, are now perceived in their ineluctable historicity; and the putative observer becomes fatally de-centered, disconnected from any epistemic link to a space-time point that can no longer be defined by geometry alone.")

Practitioners of "science studies" were caught with egg on their faces when Sokal exposed his hoax in an article in the academic journal *Lingua Franca.*[7] Clearly, the new cultural critics of science haven't a clue about what science is—or isn't.

What is most unfortunate about this whole business is that there is indeed a problem with science in our culture that needs desperately to be addressed by intellectuals—namely, the appalling split between our way of knowing and our way of believing. Even if it were not impossible, it would be foolish to abandon science as the one intellectual instrument that has given us some measure of control (for good and ill) over the natural environment. What is important is that we use our knowledge for good, that we hold fast to the sacredness of nature, that we create on this small planet a society that transcends boundaries of race, ethnicity, and gender. The alternative, already rampant about us, is a return to irrationality and both higher and lower superstition.

Paying Dues

Some of the new criticism of science has come from inside the scientific community and is informed by a thorough understanding of scientific process. As such, it is especially welcome and useful. For example, Dai Rees, secretary and chief executive of Britain's Medical Research Council, writing in *Nature*, makes a startling claim: Science has "contributed massively to human misery" by undermining traditional stable societies without offering any compensating vision of what human life might be. It is time for scientists to pay their dues, he insists.[8]

The sins of science are threefold, says Rees: two of commission and one of omission.

First, scientific discoveries are inevitably followed by technological innovations that profoundly disrupt prevailing patterns of life. For example, Rees blames technology for the teeming shantytowns of Third World cities and the high-rise housing projects of developed countries. Overpopulation, crime, disease, and mindless violence thrive in these inhumane environments, he says, and community support of individuals is almost nonexistent. Rees points also to nuclear arsenals, the accident at Chernobyl, and the destruction of rain forests as examples of technology run amok. No! respond scientists. These excesses of technology cannot be laid at the door of science: Unimaginative government planners are responsible for inhumane housing; the military-industrial complex creates weapons of mass destruction; bureaucrats caused the Chernobyl disaster; commercial adventurers destroy rain forests. Rees thinks these denials of responsibility by scientists are disingenuous. The exploitation of scientific discoveries follows a blind and remorseless logic, he claims. Technological progress follows, regardless of consequences. Like an acorn planted beneath a building, a scientific discovery grows upward whatever the cost to the edifice above. Scientists must accept responsibility for the application of their discoveries—for good or ill.

Second, Rees claims that scientific discoveries have undermined traditional belief systems that evolved over millennia. The result is a retreat by a large segment of the world population into reactionary fundamentalism, and rudderless spiritual malaise on the part of the rest. Rees might have used as an example religion in his own country. The majority Anglican communion in Britain has sought some measure of accommodation with scientific skepticism, while the minority Roman Catholic faith has more earnestly resisted the scientific ethos. The result is that there are now more churchgoing Roman Catholics in Britain than Anglicans. Again, scientists respond by saying that the search for scientific truth cannot be suppressed, nor can it be held arbitrarily in check by religious orthodoxy, and one way or another society will have to accommodate what scientists learn about the world. Easier said than done, says Rees, and if it is going to happen, scientists must show the way.

The third sin of science, according to Rees, is

a failure to make the priorities of scientific inquiry sufficiently sensitive to society's needs. For example, the billions spent on the search for the fundamental constituents of matter—particles that when forced into existence survive for only a tiny fraction of a second—would be better spent improving the day-to-day lives of ordinary people. Curiosity is a lofty human goal, reply scientists, and scientific discovery is a sublime form of creativity that is worth economic sacrifice; to parsimoniously shackle inquiring minds is to repress that which makes us most human. Rees finds merit in this argument but insists that scientists should not expect a blank check from society to pursue goals that do not tangibly contribute to the common good.

Rees's own Medical Research Council proposes that science should focus on "distinct and coherent issues central to human life," specifically:

- the planet Earth, its microscopic constitution and place in the universe, its condition and physical resources;
- the biosphere, its health and productivity;
- human health;
- products and processes;
- human and social potential.

Unfortunately, these categories are so vague as to be meaningless, and plausibly include every kind of scientific research done now or in the past. The scientists who developed the atomic bomb surely felt they were making a contribution to human and social potential by more quickly ending a globally destructive war. It is hard to imagine any research that could not be justified under one or another of these rubrics. Nor will a lot of breast-beating about shared responsibility for the excesses of technology move us forward. Albert Einstein could not have foreseen in 1915 (when he proposed the equivalence of matter and energy) the tragic accident at Chernobyl. Surely Rees does not suggest that Einstein should have turned back from his discovery or suppressed his astonishing insight into the way the world works. Does Einstein bear personal moral responsibility for the bombing of Hiroshima and Nagasaki, as well as the nuclear accidents at Three Mile Island and Chernobyl, because he discovered a fundamental potentiality of nature, the secret of the energy of stars, the font of life?

The ultimate prize for which we should all work is a science that is integrated into society, says Rees, and he is right. However, the conflict is not between science and society, as such, but between the two segments of society which I have labeled Skeptics and True Believers. The former embrace science as the most reliable guide to truth but frequently lack the moral compass and spiritual sensitivity to ensure the humane application of technology. The latter are often implacably antagonistic toward science but tend to uncritically embrace scientific technologies when it suits their purpose.

Certainly, it is not the sole responsibility of scientists to show the way to an accommodation of empirical knowing and spiritual longing. This is a task that must occupy scientists, philosophers, theologians, poets, and artists. It will mean, at the very least, that science as a way of knowing is subsumed by religion, so that the world revealed by science is perceived with an abiding sense of the sacred, celebrated liturgically, and lovingly cherished. Meanwhile, scientists muddle forward, driven by the same noble and ignoble motives that drive all human endeavors, vaguely conscious of their responsibilities toward society, but uncertain to whom or how they must pay their dues.

TEN

In Search of the Self

From all these discoveries, each of which plunges him a little deeper into the ocean of energy, the mystic derives an unalloyed delight, and his thirst for them is unquenchable.

Pierre Teilhard de Chardin

THE GREEK PHILOSOPHER Heraclitus called the soul "a spark of the substance of the stars." Modern scientists couldn't say it better. We *are* made of star stuff. Every atom in our bodies, except the hydrogen, was forged by nuclear fusion in stars that lived and died in the Milky Way galaxy before the Sun and Earth were born. According to present theories, the universe in the immediate aftermath of the Big Bang consisted only of hydrogen and helium—sufficient to make stars and gassy Jupiterlike planets, but no Earths, no carbon-based organic compounds, no life. As stars burn, they fuse hydrogen and helium into the heavier elements. When massive, short-lived stars die as supernovas, they fuse yet more heavy elements and blast these, along with some of the elements fused during their lifetimes, into space to become part of the interstellar nebulas out of which new stars are born. As more and more generations of stars live and die, the quantity of heavy elements—carbon, nitrogen, oxygen, and so on—increases in interstellar nebulas. By the time our own solar system was condensed by gravity about 5 billion years ago, the nebula out of which it formed contained about 2 percent of heavy elements, the detritus of earlier generations of stars.

Comets, which many astronomers believe are typical of the materials of the early solar system, bear within their volatile substance many of the chemical building blocks of life: organic compounds, ices, amino acids. Some of these materials were gathered by gravity into the planet Earth and became ultimately the stuff of life. We are the effervescence of creation, spume cast from the sea of matter onto a welcoming shore—carbon, hydrogen, nitrogen, and oxygen tossed into chains and streamers, lashed into animation.

Chemically, we are no different from the rest of

creation. Our elements are among the most common in the universe—a few pennies' worth of atoms. Yet we are anything but commonplace. Each of us is intensely conscious of being unique. Out of ordinary substance, a self has somehow arisen. There is no greater mystery than this: the genesis of self. How are we to understand it? The traditional answer is that God made us. We have imagined an act of creation like that of the potter who shapes a vessel on the wheel: "God fashioned man of dust from the soil, then he breathed into his nostrils the breath of life." (Gen. 2:7) It is a lovely story, but in its charming anthropomorphism vastly underestimates the power and process of God.

The new story of the self encompasses the stretchy space and time of the galaxies: atoms forged in the furnaces of stars, blasted into space in supernovas, gathered by gravity, animated on the early Earth, spun on the never-still spindles of RNA and DNA, refined across eons by the generative agency of evolution. In all of this we glimpse a divinity that is strange yet familiar, distant yet intimate, unreachable yet always present—a divinity that has been widely celebrated in creation-based mystical traditions of the world's religions, but often dismissed in our fondness for warm and fuzzy cosmologies.

When I was an undergraduate at Notre Dame University in the 1950s, I was taught that the self that inhabits my few pennies' worth of atoms is an immortal soul. It came into existence at the moment of conception by divine fiat, a fully realized human self temporarily lodged in a tiny cluster of fissioning cells, and it maintains its integrity when the body turns to dust. My theology and philosophy professors never made it exactly clear what *was* this thing, the soul. I imagined it to be a kind of airy, fairy ghost-self, much like the visible self, but immaterial, like those wispy transparent selves that rise out of dead bodies in animated cartoons. Today, after more than four decades of studying and writing about science, I am rather more skeptical about the airy, fairy soul. It is the *empirical* self that I'm interested in—the self with three *observable* characteristics: individual physical attributes, a unique ensemble of memories, and consciousness.

Explaining the first of these attributes of self was the premier scientific achievement of the twentieth century. It turns out that my physical self (*Homo sapiens*, white skin, brown eyes, black hair, a tendency to baldness in middle age, etc.) was there at the beginning, contributed by my parents, encoded in the DNA of the fertilized egg that would become me. Biologist Francis Crick, the codiscoverer of the double-helix structure of DNA, once estimated that if the biochemical information encoded on a single individual's DNA were transcribed into 500-page books, more than 1,000 volumes would be required.

As we have seen, this information is stored as a

"four-letter" code (four chemical units distributed in pairs along the double helix). One segment of paired letters is the gene for sex, another for the color of the eyes, yet another for a particular inherited disease. We are learning to read the code and to change it! (This last development is fraught with the possibility of mischief, a powerful reason why scientific knowledge and religion's sense of the sacredness of life must be quickly brought into harmony.) Geneticists have embarked upon an ambitious program to provide a complete transcription of human DNA. Most of human DNA is common to all members of our species, and great chunks of it appear to be superfluous or redundant. Only tiny bits of the code account for our individual differences blue eyes instead of brown, and so forth. Completing the transcription of the human genome will be a staggering task, but it is proceeding apace and accelerating; we can expect completion by the year 2000. The goal is no less than a complete biochemical blueprint for making a human being.

In the parlance of computers, we can now account (in principle, at least) for the self's *hardware*. What of the other two aspects of self—memory and consciousness? Clearly, stored experiences and self-awareness were not there at the beginning in the DNA. Are these things also biochemical in nature? Or might they be the defining qualities of that airy, fairy thing called the soul?

As I write, memories of my undergraduate years come to mind: walking to class in the snow with my slide rule slapping against my thigh, engineering labs, Tuesday-night physics tests, pep rallies, Saturday-afternoon football games, dancing in the darkest corners of the student center, white buck shoes, chlorophyll toothpaste, pink knit ties. Where does this stuff come from, these memories of the past collected in the attic of the mind, this ensemble of remembered experiences that is partly shared with others but in its entirety is uniquely my own? Only the most obdurate mind-body dualists doubt that memories are stored physically in that vast electrochemical network called the brain. Memories can be surgically excised or electrically stimulated. Twiddle with the *stuff* of the brain, and you twiddle memories.

Several decades ago it was widely believed that memories might be stored as RNA or protein molecules that have somehow been modified by experience, in much the same way that DNA stores genetic information. More recent research suggests that memories are recorded in the brain as networks or *traces* of interconnected nerve cells (neurons). According to this view, experience modifies the connections between neurons (synapses), creating a different trace of interconnections for each memory. There are as many as 100 billion nerve cells in the human brain, and each cell is in communication through a treelike array of synapses with thousands of other cells. Memory, then, must be imagined as a cobwebby scribble of neural circuits of the brain, much as data are stored in the interlinked circuits of a computer. If Sony can pack hundreds of the greatest works of literature onto a three-inch disc, and retrieve them at random with a handheld Discman, then it is not hard to imagine how the *far more* capacious circuitry of the human brain might record the experiences of a lifetime. Researchers are struggling to understand the physical process by which the connections between neurons are changed by experience, how short-term memories are converted into long-term memories, how trace networks stay active and distinct for a lifetime, and how memories are reclaimed and restored to consciousness. We have much yet to learn, but the broad outlines of the answers are clear: Memory is electrochemical.

Which leaves us with consciousness, this thing that seems so clearly not a *thing*, so intensely real and yet unknown, so palpably unpalpable. Is consciousness also made of star stuff—atoms of hydrogen, carbon, oxygen, and nitrogen engaged in a dervish electrochemical dance? Or will consciousness forever escape the grasping theories of the explainers into unmeasurable, immortal spirit—the "soul" of Theology 101, circa 1954. Sometime in my third year of college, as required reading for a course called Junior Seminar, I encountered Carl Jung's *Modern Man in*

Search of a Soul. What most impressed me at the time was the title of the book, which implied that the soul might be something we had to look for. Why should that be so? I thought. After all, the soul was the same as consciousness, and consciousness was simply there, immediately experienced—the divinely created, immaterial self. It required no explanation; it was *beyond* explication, off-limits for scientific analysis. The supernatural soul would not be found in the lab, nor reveal itself under the autopsist's knife. Or so I had been taught. But scientific curiosity, I subsequently discovered, resists arbitrary theological or philosophical limits on its quest for understanding, and eschews any supposed body-spirit dualism that would place consciousness beyond the pale of rational explanation. The soul is fair game for science. A formidable quarry, but fair game.

Nevertheless, consciousness has so far resisted scientific understanding. Of the three presumptive elements of self—physical uniqueness, ensemble of memories, conscious awareness—consciousness is the last recalcitrant mystery. The prevailing school of thought among cognitive scientists holds that consciousness is biochemical in nature and ultimately amenable to naturalistic explanation. Indeed (so the story goes), if computers could be constructed with the same level of complexity as the human brain (100 billion massively interconnected elements) and with sufficiently rich sensory inputs, then we should not be surprised to encounter conscious machines. Of course, no computer yet remotely approaches the complexity of the human brain. Simply to write out the numeral corresponding to the number of distinct possible modes of connection of the 100 billion neurons of the human brain (a number written exponentially as ten to the 100 trillion power) would require 100 million 500-page volumes.

When we consider the astonishing power and versatility of contemporary computers, which in terms of complexity are to the human brain as a skateboard is to a Ferrari, a biochemical basis for consciousness seems eminently reasonable. Perhaps. The case is not proven. It would be foolish, however, to stake all against it, as do Appleyard, Havel, and a host of other modern-day obscurantists, for the proof will be in the pudding. Massively parallel, ultra-high-speed computers, modeled on the molecular machinery of the human brain and comparable to the brain in complexity, await us in the next century. If machine consciousness turns out to be indistinguishable in certain essential aspects from human consciousness, then Junior Seminars of the future can pose the fascinating question: Do machines have souls?

A Computer Made of Meat?

True Believing fundamentalists and many academic critics of science resist the idea that minds are merely computers made of meat (to use a provocative phrase of Marvin Minsky, a pioneer researcher in artificial intelligence). Our minds, they say, are free, creative, and intuitive, qualities that are forever beyond the capability of any conceivable computer. Cognitive scientists respond by saying: Human freedom, creativity, and intuition are high-level abstractions describing the behavior of hugely complex biochemical systems running programs that are partly hardwired by the genes and partly sensitive to the rich variety of sensory experience. Computers with a comparable level of complexity and an equally variegated sensory input might also be free, creative, and intuitive. Ockham's razor would suggest that we not invoke a disembodied explanation for consciousness until we are required to do so; so far, cognitive science has done very well without it.

The biologist E. O. Wilson writes: "There have always been two kinds of scientists, two kinds of natural philosophers. The first look upon the Creator, or at least the ineffability of the human spirit, as the ultimate explanation of first choice. The second follow the venerable dictum attributed to Polybius that, whenever it is possible to find out the cause of what is happening, one should not have recourse to the gods."[1] In the study of consciousness, the gods are in retreat. Time and progress have a way of turning supernaturalist True Believers into Polybian Skeptics.

So is the mind a computer made of meat? Many

people find such a notion disheartening. We have been so long immersed in dualities of body and mind, matter and spirit, natural and supernatural, that we resist any attempt to reduce our true selves to body, matter, and nature. We are frightened by the reduction of consciousness to biochemistry, even though such a reduction will have not the least impact upon our day-to-day lives—our loves, hates, elations, fears, joys, sorrows. Is there an alternative to biochemical theories of consciousness short of invoking the immaterial soul of Cartesian dualism? A number of recent authors suggest that the iffy probabilities of quantum physics provide an escape from the slavish determinism of computers. Their books—with titles like *Quantum Psychology*, *The Quantum Self*, and *Star Wave*—have become the scriptures for a New Age religion (call it quantum mysticism) that has sprung up on the fringes of science.

Quantum physics describes the behavior of the world on the subatomic scale, and the laws of quantum physics are odd indeed, even spookily ineffable. Things happen in quantum physics according to probabilities; no single atomic event is precisely determined in the classical Newtonian sense. An electron might be here or it might be there; only a probability can be known in advance and only observation will tell. Some recent experiments suggest that a particle might be *in two places at the same time*. Might not this hazy indeterminism of the quantum world provide a *physical* basis for human freedom and creativity?

Well, maybe. First of all, it is important to recognize that the so-called uncertainties encountered on the quantum scale of nature are themselves exactly described by mathematical laws that have been repeatedly confirmed to a high degree of precision. As Gross and Levitt point out in *Higher Superstition*, the Heisenberg Uncertainty Principle "is not some brooding metaphysical dictum about the Knower versus the Known, but rather a straightforward statement, mathematically quite simple, concerning the way in which the statistical outcomes of repeated observations of various phenomena must be interrelated."[2] Attempts by New Age gurus to turn quantum physics into a religion of magic, miracles, and mysticism are profoundly misguided. Frank Tipler's attempt to confirm the resurrection of the body with quantum statistics is simply silly. So far, no quantum basis for human creativity and freedom has been demonstrated, and most popular books on the subject are no more than wishful longing for transcendence posing as physics.

One book on this topic that deserves serious consideration is Roger Penrose's *The Emperor's New Mind*, a solidly scientific plea for a nonmechanistic theory of consciousness. Penrose is a brilliant mathematician and physicist, a collaborator with Stephen Hawking on the physics of the universe's origin. Somewhere deep in the brain, suggests Penrose, unpredictable quantum events—the twitches of individual electrons—trigger conscious thoughts by a kind of chain reaction. An electron skips in a totally random way, setting off a cascade of electrochemical events that ultimately manifests itself as conscious thought. Such thoughts partake of the undetermined character of the random events that triggered them; hence, they are unpredictable by any present scientific theory.[3]

The theory is deceptively attractive. According to Penrose, freedom, creativity, and intuition bubble up from the quantum deep. Alas, nowhere in his 400-page book does he offer experimental evidence for a quantum basis for consciousness. It is unlikely that neurons in the human brain can be triggered (made to fire) by the kind of twitchy electronic activity that is described by quantum physics. Certain energy thresholds must be met before brain cells are activated, and many separate electronic stimulations are needed to reach those thresholds, statistically washing out any role for quantum physics. An analogy: Role a die once, and you have a one-in-six chance of predicting the outcome; that's the sort of iffy prediction provided by quantum physics. Roll a die a million times, and you can be sure that the total will add up to 3.5 million, or close enough to

3.5 million to make operating a casino a sure bet. That's the kind of certainty you meet in classical physics.

Quantum randomness at the level of individual electron events is washed out on the macroscopic scale. If quantum effects were important in the brain, then long-term memory, which requires a high degree of macroscopic stability, would be problematic at best. However, Penrose is a bright fellow, and we should keep an open mind. He is convinced that the mind will never be emulated by a computer of the sort we use today, and he may be right, but so far his intuition has no more empirical basis than the starry-eyed ruminations of the quantum mystics.

So where does that leave us? Waiting. But not idly. Physicists, psychologists, biologists, neuroscientists, artificial intelligence researchers, computer scientists, and microchip fabricators are vigorously engaged in what will surely be the premier scientific quest of the twenty-first century: understanding the mind. My intuition in these matters is that there is nothing mystical (or even quantum physical) about consciousness, and that a computer of sufficient complexity and with rich sensory input might one day be fully conscious. (But *not*, of course, fully human; our bodies of flesh and blood are an intimate and essential part of our beings.) If one believes that the Cartesian duality of matter and spirit is a bankrupt notion in the late twentieth century—as almost all scientists do—then the previous statement is almost a truism. The mind *is* the brain; the brain *is* an electrochemical mechanism; therefore the mind is an electrochemical machine. But I suspect we have a few theoretical surprises in store before a final understanding is achieved.

All Things Flow

If we are frightened by the idea that the self can be reduced to matter and mechanism, then perhaps it is because we have outdated notions of what those concepts mean. Matter, in modern physics, has become a thing of marvelous subtlety and potentiality. Long ago physicists gave up the idea of atoms as chunks of hard stuff writ small. As we probe the structure of matter, we find ourselves awash in a sea of cosmic music—surging, billowing, animating, never at rest. The real constitution of things is accustomed to hiding itself, said Heraclitus, who also told us that we cannot step in the same river twice. Matter flows like an invisible flame, brightening and energizing the universe. Step in the river and burn your foot. Step in the river twice, and it is neither the same river nor the same foot. Matter, in the view of contemporary physics, is everything we had hoped spirit to be, and more. Mechanism, meanwhile, has come to mean much more than a Renaissance clockwork or windup toy. The new metaphor for mechanism is drawn from the concept of ecological wholeness: Every part of a complex system plays a role in the evolution of the whole.

To admit that we are matter and mechanism is to ground our selves in the wholeness of the cosmos. In the new physics, self coalesces from the stuff of the stars, exists briefly (as the river and the foot exist briefly), then flows back into wholeness. Such a concept of self can be ennobling, cosmic, ecological—more so than the ghostly spirit soul I encountered in freshman theology.

Our concept of self changes from generation to generation, and is shaped by the cosmology of the times. A hundred years from now we will know more about the world than we know today, and our concept of self will have evolved to become fuller, richer. Scientific knowledge is cumulative, advancing toward more complete understanding. It is unfortunate that many philosophers, theologians, and social critics do not let their speculations be informed by the extraordinary things we are learning about molecular biology and the neural basis of consciousness. To admit that the mind is electrochemical does not diminish our concept of self; rather, it suggests that the cosmos was charged with the possibility of becoming conscious from the first instant of creation. The newly emerging concept of self is materialistic and mechanistic; it is also capacious enough to embrace not only the future but also the past, and expansive enough to entangle the self with the rest of creation.

In an article in *Studies in the Spirituality of Jesuits*, the Jesuit priest David Toolan urges his fellow theologians to embrace a new kind of ecological theology, one that is informed and energized by what science has learned about the universe: "As observers . . . we are not structurally different from the crystal, the plant, the animal, the order of the world that we observe. We understand them because of our affinity: Like them, both determined and riddled by chance, we are noise, disorder and chaos on one side; complexity, arrangement and order on the other. Like the rest of the cosmos, we give signs. In a vast chain of conversions of chancy energy, we are simply the final alchemists, the last transformers and interpreters, the ultimate black box."[4]

The American poet Mary Oliver says something similar:

> *Is the soul solid, like iron?*
> *Or is it tender and breakable, like*
> *the wings of a moth in the beak of the owl?*
> *Who has it, and who doesn't?*
> *I keep looking around me.*
> *The face of the moose is as sad*
> *as the face of Jesus.*

The swan opens her white wings slowly.
In the fall, the black bear carries leaves into the
darkness.
One question leads to another.
Does it have a shape? Like an iceberg?
Like the eye of a hummingbird?
Does it have one lung, like the snake and the scallop?
Why should I have it, and not the anteater
who loves her children?
Why should I have it, and not the camel?
Come to think of it, what about the maple trees?
What about the blue iris?
What about all the little stones, sitting alone in the
moonlight?
What about roses, and lemons, and their shining leaves?
What about the grass?[25]

To understand that we are structurally no different from the rest of the cosmos is to let ourselves expand into infinity. It remains to be seen if or how theologians will reconcile the traditional notion of immortality with the new materialistic, mechanistic self. It may be best to simply toss immortality aside and get on with the celebration of the cosmically embedded self.

We have more to gain than to lose. In the biological and neurological discoveries of the past quarter century, we catch a glimpse of a marvelous cosmic self in resonance with the universe, a distillation of stellar fires, a Heraclitean spark of the substance of the stars. The Hebrew and Christian scriptures tell us that God created the first man and woman out of the slime of the Earth, breathed life into them, and pronounced them good. The myth is consistent with our current understanding of the nature of life. According to the best scientific theories, we are literally animated slime. Now we must relearn to think ourselves "good."

ELEVEN

Knowing the Mind of God

The more we learn, the more we are—or ought to be—dumbfounded. . . . [O]ur proper business is to learn more and more and thereby separate our mere ignorance from genuine mystery.

Lewis Thomas

RELIGIOUS TRUE BELIEVERS stubbornly disregard the cosmological contradictions embedded in their faith. Meanwhile, some scientists, particularly physicists, with equal stubbornness, claim that scientific theories can tell us the ultimate meaning of the universe. This latter view—that knowledge can be made complete and mystery extinguished—retains its hold on a certain sort of scientist, although history has repeatedly debunked any closure on research.

In 1894, on the eve of one of history's most prolific eras of scientific discovery, the great American physicist Albert Michelson said: "While it is never safe to affirm that the future of Physical Science has no marvels in store even more astonishing than those of the past, it seems probable that most of the grand underlying principles have been firmly established and that further advances are to be sought chiefly . . . in the sixth place of decimals." In other words, we know everything that is worth knowing; all that is left for scientists to do is tidy up along the edges. Another important late-nineteenth-century physicist, Lord Kelvin, affirmed outright that science was at an end. Only two little clouds remained on the horizon which physics had not yet satisfactorily accounted for, he said: the energy spectra of black body radiation and the negative result of the Michelson-Morley experiment. Ironically, it was precisely these two "little clouds" that gave rise to the quantum and relativity revolutions, opening grand new vistas for physicists and putting the kibosh on Kelvin's overweening finality.

Now, a century later and the lesson still not learned, distinguished physicist Stephen Hawking concludes his best-selling book *A Brief History of Time* with the suggestion that we might be on the threshold of discovering an ultimate theory of the universe that would explain why it is that we and the universe exist. "If we find the answer to that," he says, "it would be the ultimate triumph of human reason —for then we would know the mind of God."[1]

The Grand Design

It is worth noting that both Michelson and Hawking hedged their bets with words like *never safe to affirm, probable, maybe*, and *if*, characteristic language of Skeptics. Nevertheless, both men assert confidence in the power of the human mind to grasp ultimate reality. Hawking even predicts access to the mind of God! Great physicists don't necessarily make reliable theologians and philosophers.

Readers who come to Hawking's book looking for a personal encounter with God's mind will be disappointed. Hawking's invocation of the deity most often records his absence. Indeed, Hawking's own labors as a theoretical physicist would seem to restrict ever more severely any actual or potential role for a cosmic being of the traditional world-meddling sort. Isaac Newton, Hawking's illustrious predecessor as Lucasian Professor at Cambridge University, confined God's role in the universe to that of a Great Clockmaker who set things going and then retired from his creation. Hawking proposes a universe that has no boundary in space or beginning in time, thereby removing even the need for a Clockmaker.

The essence of Hawking's natural theology is this: God's mind is identical with a plan of creation of stunning simplicity and generality, the so-called Grand Unified Theory (GUT) sought by late-twentieth-century physicists. He believes we have caught many essential elements of that plan and now stand at the threshold of the Grand Design itself.

Hawking's engaging invitation to know God's mind is a bait-and-switch come-on, for God's thoughts, as revealed by Hawking, turn out to be couched in the inscrutable language of mathematical physics. Few of us are capable of reading the Grand Design in the language in which it is written. The translation into ordinary English, as, for example, in Hawking's best-selling book, means something essential has been lost. This relative inaccessibility of modern cosmological thought is surely one of the reasons for the stubborn atavism of fundamentalists, and the rampant popular interest in pseudosciences and New Age quackeries that offer an aura of science without the drudgery. Most readers of *A Brief History of Time* probably give up somewhere about page twenty-five, when the discussion veers out of familiar territory into non-Euclidean space-time. The promise of an easy path to God's mind soon gives way to a mathematical thicket of superstrings and black holes.

It is not God's mind that is being revealed here, but Hawking's. This in itself makes the book worth reading, because Hawking's mind is surely one of the most brilliant of our century. Moreover, the universe that has been revealed by Hawking's efforts as a theoretical cosmologist is an apt stimulus for wonderment. But "the mind of God"? Not likely. It is the nature of God to reside in mystery—ineluctable, inexhaustible mystery. And we do not need to understand the cabala of mathematical physics to apprehend the *mysterium tremendum.* We need only look out the window.

Dreams of Final Theory

In a song called "You and I," songwriter Meredith Willson suggests that only lovers know why the sky is blue and why birds sing. Nobel Prize–winning physicist Steven Weinberg thinks he knows too. The sky is blue because air molecules scatter (bounce about) short-wavelength blue light more effectively than the other colors of the spectrum, so the Sun's blue light seems to come to us from every part of the sky. And the singing of birds can be explained by molecular DNA. At a deeper level, these molecular explanations can be explained by the laws of quantum physics. And at a yet deeper level, there are a few primal laws of nature that we have begun to glimpse, the Grand Unified Theory, that bind everything in a satisfying wholeness. In fact, says Weinberg, everything in the universe, including the universe itself, will ultimately be understood in terms of a "final theory" of stunning simplicity. It will be a mathematical theory, derived from the experimental investigations of high-energy particle physicists. It will be the glittering point where all lines of explanation converge.

This is the theme of Weinberg's book *Dreams of a Final Theory*. His reductionist optimism is the same as that of Michelson, Kelvin, Hawking, and others afflicted with the physicist's occasional unbridled hubris. The singing of birds is biology, they say; biology is chemistry; chemistry is physics; physics is elementary particle physics; and behind elementary particle physics is the GUT of the universe. Weinberg writes: "Reductionism is not a guideline for research programs, but an attitude toward nature itself. It is nothing more than the perception that scientific principles are the way they are because of deeper scientific principles (and, in some cases, historical accidents), and that all these principles can be traced to one simple set of connected laws."[2]

This brand of strong reductionism has a nasty odor about it. The idea that we are just a bunch of elementary particles bouncing about in the void, guided by deterministic laws that allow no variation, feels chillingly impersonal. If there is an overall plan in the world, we would like it to be something more than the mathematical blueprints of Hawking and

Weinberg, something to do, maybe, with blue summer skies and bird songs and you and me.

Strong reductionism of Weinberg's sort teeters toward True Belief. Fortunately, Weinberg offers his views tentatively, as well a physicist should. He calls his chapter on all of this "Two Cheers for Reductionism," not the traditional three cheers, modestly hedging his bet. And he admits "historical accidents" into his definition of how the world works, although only as a parenthetical phrase, and a qualified parenthetical phrase at that. But though he makes the requisite nod to scientific skepticism, his confidence in finding a final theory is unassailable. The reductionist way of explaining the world must be accepted, he insists, not because it makes us feel good but because that's the way the world works. Elementary particle physicists have every right to claim their discipline is more fundamental than other ways of knowing because it *is* more fundamental, he says. Period.

And he makes a compelling case. Elementary particle physics is without doubt one of the most stunningly successful inventions of the human mind, guiding our imaginations into the vibrant heart of matter. But chilling and impersonal, yes, and only tenuously connected to the world in which we live our lives. When all is said and done, I suspect that songwriter Willson can teach us as much as Weinberg about why the sky is blue and why birds sing. Strong reductionism is not science. It's a personal philosophy, a matter of faith. And the person who believes that the melodies of birds will not, even in principle, be ultimately explained by elementary particle physics has as much or as little basis for his belief as the reductionist.

I would turn Weinberg's definition on its ear: Reductionism has been a wildly successful guideline for research, but as an attitude toward nature it's a flop. We have learned much about the world by breaking things down into their component parts, but much remains unexplained, including most of the things that touch upon our day-to-day lives. Who knows what we might learn by adopting a less reductionistic approach? I'm not talking about "gosh-isn't-it-all-wonderful" New Age holism, but rather new kinds of mathematical and empirical synthetic science.

Reductionism has worked well so far because it's the only kind of science we have been able to handle mathematically, and mathematics is the *fons et origo* of our success. However, with the advent of superpowerful, high-speed computers, synthetic ways of explaining things may emerge which are as quantitative *and* as successful as the kind of reductionist science we have been doing since Galileo; indeed, computers have already begun to transform the way science is done.

What is missing in the reductionist program is

historical accident, which as you will recall Weinberg made a place for, but only parenthetically. Complex systems, such as life and consciousness, are notoriously susceptible to chance, and strong reductionism takes little note of contingency. My own guess is that life and consciousness are emergent phenomena that arise when material systems have reached a certain threshold level of complexity, and only a computer-based science of complex systems, seasoned with a dash of historical accident, will explain them. Reductionism has been enormously successful at expanding the terrain of our knowledge, but mystery still laps our ever-lengthening shore. By the end of the next century, we may look back upon reductionist physics as hopelessly naive.

Biologists and cognitive scientists tend to be less reductionist than physicists; they presumably have more respect for the irreducible potentialities of complex systems. Harvard biologist E. O. Wilson writes: "Our sense of wonder grows exponentially: the greater the knowledge, the deeper the mystery. This catalytic reaction, seemingly an inborn human trait, draws us perpetually forward in a search for new places and new life. Nature is to be mastered, but (we hope) never completely. A quiet passion burns, not for total control but for the sensation of constant advance."[3]

Wilson's modesty is a refreshing contrast to Weinberg's hubris. A little hubris is not necessarily a bad thing for a scientist; anyone who would attempt to explain the universe must possess some measure of presumption. The important thing is not to let hubris get out of control. As Skeptics, both Steven Weinberg and E. O. Wilson qualify their positions parenthetically. Here's my take on strong reductionism, which like Weinberg's opinion is a matter of faith: *No theory conceived by the human mind will ever be final. The universe is vast, marvelous, and deep beyond our knowing; its horizons will always recede before our advance. All dreams of finality are (probably) futile.*

Accelerators and Cathedrals

It is a common conceit of high-energy particle physicists to compare their giant accelerating machines to the Gothic cathedrals of the Middle Ages. Robert Wilson, former director of the Fermi National Accelerator Laboratory (Fermilab), draws these analogies: Cathedrals were intended to reach new ultimates of height, and accelerators push new limits of energy; the aesthetic appeal of both structures is based on cutting-edge technologies; the builders of cathedrals and accelerators were daring innovators, fiercely competing along national lines, yet basically internationalists. Leon Lederman, Wilson's successor at Fermilab, adds rather grandly: "Both cathedrals and accelerators are built at great expense as matters of faith. Both provide spiritual uplift, transcendence, and, prayerfully, revelation."[4]

What physicists are looking for with their accelerators are the ultimate constituents of matter, the first particles to condense from the primordial radiation of the Big Bang, particles that can exist only at exceedingly high energies. High-energy physicists smash protons against protons, electrons against electrons, and protons and electrons against atomic nuclei with powerful machines such as those at Fermilab in Illinois and CERN in Europe. Out of these titanic microcollisions comes a bewildering spray of new particles: pions, muons, neutrinos, W and Z particles—the list is endless. Every time physicists jack up the energy of the bombarding particles, strange new stuff comes fleetingly into existence, giving us a glimpse of the constituents and structure of matter (and insight into how the universe coalesced from the hot energy of the Big Bang). Every increase in the energy of the machines takes us closer to the conditions that prevailed during the first moments of creation.

Behind the spray of new particles, physicists believe they can deduce the existence of an ultimate particle or force, called the Higgs after the British physicist Peter Higgs, who first proposed its existence—a point where all lines of explanation converge, the "final theory." If the Higgs can be forced into existence for even the tiniest fraction of a second, we would glimpse the Creator's primal plan—or so say the physicists. There is just one hitch. With contemporary technology, creating the Higgs will require a machine of staggering size and and complexity.

Several years ago, American physicists proposed to build such a machine, an underground oval proton racetrack, fifty-two miles in circumference, called the Superconducting Super Collider, which would cost taxpayers more than $10 billion. Two beams of protons would travel in opposite directions around the ring, kept in their tracks by 3,840 magnets, each fifty-six feet long, and focused by 888 other magnets, the magnets altogether containing 41,500 tons of iron and 12,000 miles of superconducting cable cooled by 525,000 gallons of liquid helium. When the two speeding beams are caused to collide, the Higgs (if it exists and if predictions of its properties are correct) will flicker briefly into existence out of pure energy, a kind of Einsteinian transubtantiation. Thus the analogy with the Gothic cathedrals.

The cathedrals, too, were built at great public expense, and few of us would wish they had not been built. Physicists push the analogy further. The Higgs is the "Holy Grail" of science, they say with religious fervor. Leon Lederman of Fermilab calls the Higgs the "God particle." And Hawking, of course, speaks of reading the mind of God. The quest for the Higgs, say the physicists, is the same as the quest of American historian Henry Adams when he visited Chartres cathedral near the end of the last century: "The struggle of [man's] own littleness to grasp the infinite."

The analogy is breathtaking. Alas, it contains a flaw.

The Gothic cathedrals were the most prominent objects in the lives of the people who built them. They soared above townscapes. They were sites of common worship and ritual passages of birth, marriage, and death. They were scriptures in stone, accessible for all to read in a language of universal, easy-to-grasp images. A requirement of every cathedral was that it be able to accommodate the entire population of its town, from powerful lords to poorest peasants. By contrast, accelerators are built underground, out of sight. The why and how of the machines is understood only by a tiny subset of the societies that build them. What happens in the machines would appear to have nothing whatsoever to do with our daily lives.

The cathedrals were largely paid for by voluntary donations. Sometimes the bishop or cathedral chapter contributed substantially, and the local prince or lord might have given as well, but most of the money came from ordinary folks. The contributors' motives were not altogether selfless. A contribution might purchase an indulgence, a sort of redeemable spiritual coupon that would shorten the time of the donor's suffering in purgatory. In any case, the extraordinary sacrifices involved in building the cathedrals were thought to evoke God's extraordinary favors in return.

By contrast, private voluntary donations or investments play no role in building accelerators. After all, what will the Superconducting Super Collider offer us in return? Exotic particles that exist for a tiny fraction of a second, leaving behind mere blips among data stored in a computer. Confirmation of theoretical speculations that only a few hundred physicists fully understand. It is no surprise that in the end the American Congress pulled the plug on the project.

The cathedral analogy is a sham. Gothic cathedrals represented a spontaneous outpouring of love and sacrifice on the part of the people who paid for their construction. High-energy physicists have a long way to go in explaining their quest for a final theory before they can reasonably expect similar enthusiasm and generosity from taxpayers. All scientific research has a revelatory function, and the esoteric investigations of these physicists bring news of the first moments of creation, but so far the American public is having none of it. And rightly so.

The religious analogy does not yet stick. We have not yet made the link between scientific knowledge, spirituality, and celebration.

Absolute Darkness, Absolute Light

What about God? asks Steven Weinberg in the penultimate chapter of *Dreams of a Final Theory*. As we approach a final theory, and understand that everything that exists unfolds from its primordial beginnings according to simple, inevitable laws—what about God? Knowing the final laws of nature, "we would have in our possession the book of rules that governs stars and stones and everything else," he writes, with sublime self-confidence. Whatever one's religion or lack thereof, it is irresistible to speak of the final laws of nature as "the mind of God." But why do so? asks Weinberg. If we define God so broadly that he can be identified with impersonal laws of nature, then what's the point?

Weinberg quotes Einstein as believing in "Spinoza's God who reveals Himself in the orderly harmony of what exists, not in a God who concerns himself with fates and actions of human beings." But what difference does it make, asks Weinberg, to use the word *God* in place of *order* or *harmony*, except to avoid the accusation of having no God? If you want to say that "God is energy," he says, then you can find God in a lump of coal. And he adds, "But if words are to have any value to us, we ought to respect the way that they have been used historically, and we ought especially to preserve distinctions that prevent the meanings of words from merging with the meanings of other words."

Weinberg professes himself more sympathetic to fundamentalists, who at least know what they mean by *God*, than to religious liberals, who define *God* so vaguely as to have no meaning at all: "The more we refine our understanding of God to make the concept plausible, the more it seems pointless," he says. If God is the equations of particle physicists, then the word *God* has been gutted of its traditional meaning. And if the equations of particle physicists—the GUT, the Grand Unified Theory—are all that exists, then the God of the prophets and the creed is well and truly nonexistent.

In this respect, Weinberg is right. The traditional, scriptural concept of God as a personal being, interested in our individual fates and capable of intervening at will in the workings of nature, does not rest easily with what science has discovered about the creation. Weinberg's mistake is to assume that the only alternatives are the God of the scriptures, on the one hand, and the God of the GUT, on the other.

But *God* has another independent usage within the mystical tradition, a usage that is universal, nonsectarian, and inclusive, that goes back to the origins of religious observance. The Greek novelist Nikos Kazantzakis writes: "We have seen the highest circle of spiraling powers. We have named this circle God. We might have given it any other name we wished: Abyss, Mystery, Absolute Darkness, Absolute Light, Matter, Spirit, Ultimate Hope, Ultimate Despair, Silence. But we have named it God because only this name, for primordial reasons, can stir our heart profoundly. And this deeply felt emotion is indispensable if we are to touch, body with body, the dread essence beyond logic."[5]

The God of the spiraling powers resides in nature beyond all metaphors, beyond all scriptures, beyond all "final theories." It is the ground and source of our sense of wonderment, of power, of powerlessness, of light, of dark, of meaning, and of bafflement. It is the God whose history began with the first human who experienced awe, contingency, fear. It is the God of mystics of all cultures and creeds. We stand on the shore of knowledge and look out into the sea of mystery and speak his name. His name eludes all creeds and all theories of science. He is indeed the "dread essence beyond logic."

Science extends the shore along which we are able to perceive the mystery, but it does not deplete the mystery. As knowledge deepens, so does wonder. God appeared to Julian of Norwich in the form of Christ and placed in her hand a little thing the size of a hazelnut, "as round as any ball." She looked at it and wondered what it might be. He answered her: "It is all that is made." Even if the Higgs were placed in our hand, its very existence would confound our knowledge and confirm our ignorance.

The mistake of GUT-obsessed physicists is the mistake of the ontological theologians: the identification of God with knowledge rather than with mystery. Without the experiential grasp of the mysterious essence of creation, science is a mere praxis, a prop for technique. And without science, experiential union with God is limited to that part of creation we see outside our windows; the universe of the galaxies and the DNA must remain unknown.

Only at one point in his book does Weinberg lean briefly toward the mystical. He writes: "I have to admit that sometimes nature seems more beautiful than strictly necessary. Outside the window of my home office there is a hackberry tree, visited frequently by a convocation of politic birds: blue jays, yellow-throated vireos, and, loveliest of all, an occasional red cardinal. Although I understand pretty well how brightly colored feathers evolved out of a competition for mates, it is almost irresistible to imagine that all this beauty was somehow laid out for our benefit. But the God of birds and trees would have to be also the God of birth defects and cancer."

Yes, Weinberg, yes. Go that little bit further. Let your soul go free for a moment into that scene outside your window, into the vistas of cosmic space and time revealed by your physics, and there encounter, gape-jawed and silent, the God of birds and birth defects, trees and cancer, quarks and galaxies, earthquakes and supernovas—awesome, edifying, dreadful and good, more beautiful and more terrible than is strictly necessary. Let it strike you dumb with worship and fear, beyond words, beyond logic. What is it? It is everything that is.

TWELVE

The Weight of Facts

The scientist has returned to the larger culture with stories, awesome and frightening, but stories that serve to mediate ultimate reality to the larger culture.

Brian Swimme

TRADITION HAS IT that Adam was allowed by the Creator to name all the creatures of the Earth. It was a daunting task! According to biologists, there are between 10 and 100 million species of living organisms. That means if Adam thought up a name a minute for sixteen hours a day (Sundays included), it would take him between 30 and 300 years to complete the job. Still, it must have been great fun coming up with tags like "duck-billed platypus," "tufted titmouse," and "precious wentletrap" (a gastropod of Southeast Asia).

I was set to thinking about Adam's task when I came across a book called *The Common Names of North American Butterflies*, compiled by zoologist Jacqueline Miller of the University of Florida. Her purpose was to bring some order to the jumble of common names used by amateur and professional butterfly enthusiasts. The book is sheer poetry. Creamy checkerspot. Buckwheat blue. Hop-eating hairstreak. Bloody spot. Rainbow skipper. Mad flasher. Here is a list of luscious language to delight the soul of Vladimir Nabokov or James Joyce, those archmagicians of the English tongue. Nabokov was a lepidopterist of note, and his name is twice recorded in Miller's book as Nabokov's blue and Nabokov's fritillary. Joyce may never have netted a butterfly, but the author of *Ulysses* and *Finnegans Wake* would certainly have appreciated the redundant swarthy skipper and Mrs. Owen's dusky wing for their names alone. The parsnip swallowtail is also called parsleyworm, celeryworm, and carawayworm, which suggests a certain catholicity of taste on the part of the butterfly. The flying pansy is alternately the California dog face, which suggests that one lepidopterist's beautiful is another's ugly. One wonders if the lost-egg skipper misses its progeny.

Certain professional (and even amateur) lepidopterists blanch at the very mention of common names and their attendant confusion. They plump for the use of scientific nomenclature exclusively—for clarity. Adam, of course, was no scientist, so he went about his work with reckless disregard for nature's underlying order. (As a pre-Darwinian, he perhaps assumed that God created the creatures with the same whimsical abandon that he, Adam, now named them.) One would never know from their common names that the goggle eye and the red-eyed nymph butterflies are first cousins.

Our system of scientific naming has a history going back to Aristotle but owes most to the eighteenth-century Swedish botanist Carl von Linné, better known by his Latinized name, Linnaeus. He proposed a binomial system, consisting of a genus designation for all species in a closely related group, followed by a species-specific modifier. Thus, goggle eye becomes *Cercyonis pegol* and red-eyed nymph becomes *Cercyonis meadii*, with their kinship made manifest by the genus designator. To decide which plants were intrinsically related, Linnaeus looked particularly at their sexual characteristics. His successors extended his naming system to animals, minutely examining anatomical features to decide relationships. Modern molecular biologists, comparing the chemical structures of proteins and DNA, have not improved much upon the family relationships established by nineteenth-century Linnaean taxonomists.

Many years ago, I visited Linnaeus's country

house near Uppsala in Sweden. It was a charming place, surrounded by nature's bounty. Butterflies flitted in the dooryard. The interior walls were papered with marvelous drawings of plants. In this summery Eden, Linnaeus tossed out Adam's common names and proposed his system of Latin binomials. He knew that nothing is well described unless well named, and that nothing is well named until well described. Naming and exact description go hand and hand, and, if carefully done, reveal patterns of order implicit in nature itself.

The idea of an intimate connection between *naming* and *understanding* was in the air in the eighteenth century. Not long after Linnaeus proposed his nomenclature system for biology, Antoine Lavoisier set out to do much the same thing for chemistry. In the preface to his great work, *Elements of Chemistry*, Lavoisier quotes the philosopher Condillac: "We think only through the medium of words. . . . The art of reasoning is nothing more than a language well arranged." Lavoisier goes on to tell us: "Thus, while I thought myself employed only in forming a nomenclature, and while I proposed to myself nothing more than to improve the chemical language, my work transformed itself by degrees, without my being able to prevent it, into a treatise upon the Elements of Chemistry." So, too, did Linnaeus's revision of biological nomenclature lead inexorably to the work of Charles Darwin and the theory of evolution by natural selection.

Behind Linnean nomenclature stands a hugely profound idea: *There is a world out there that exists independently of ourselves*, in a space that is not human space, in a time that is not human time. The change from goggle eye to *Cercyonis pegol* represents a continental divide in human thought. Until the Scientific Revolution of the seventeenth century, meaning flowed from ourselves into the world; afterward, meaning flowed from the world to us. Until the Scientific Revolution, the universe and all of its creatures were assumed to have been created ex nihilo by God as arena and supporting cast for the human drama of salvation. "All the world's a stage," said Shakespeare, and he meant it quite literally. The abode of man was believed to be the fixed center of the cosmos, attended by planets and stars. Adam (and Eve, of course) stood on the top material rung of a chain of being that stretched from the dreggy center of the Earth to the foot of God's throne. Since Adam was the chief of God's corporeal creatures, and raison d'être for all the others, it was only proper that he should be given the task of naming. Adam and his descendants enjoyed supreme dominion.

Then the Scientific Revolution gave expression to a radical new idea, an idea that had been perking away in the background of human thought since the

Greeks, perhaps the most revolutionary idea in the history of human thought: *The cosmos exists independently of ourselves.* We are small, contingent parts of something that existed long before we appeared on the scene. Human life could vanish from the planet, and the cosmos will continue. We are as incidental to the cosmos as are ephemeral mayflies to the planet Earth.

At first glance, this was shattering news. Indeed, the majority of us have not yet come to terms with it. We reject science as an instrument of revelation because we are made uncomfortable by what science reveals: *Our lives are brief, our fate is oblivion.* But there is an upside to this new knowledge, if we are smart enough to perceive it. Once we admit that the world exists in its own right, it becomes possible to know the world on its own terms—or at least we can *try* to know the world on its own terms. In doing so, we discern a divinity who is not merely a projection of our own hopes and fears, a gray-bearded extrapolation of ourselves. Instead of making the world a mirror for ourselves, we make our minds mirrors for the world. This is the task that separates us from the beasts, our highest calling: to become *knowers* of creation. When the world *as it is* is humbly, skeptically ensconced in our minds and hearts, we will have made ourselves the instruments of the universe's self-reflection.

Linnaeus proposed his system of binomials, and

Lavoisier his chemical nomenclature, to bring us closer to the world as it is. This is also the reason we adopt mathematics as the premier language of science. "Well-arranged" languages are nets we throw to catch the world; with them we snare the universe of the galaxies and the DNA, a universe we might never have known had we remained confined by the messy language of ordinary discourse.

This is the grandest adventure of all: To cut loose from our secure moorings and launch ourselves into the space and time of the galaxies. Risky business, of course. A danger of vertigo. A fear of infinite dilution. Oblivion. But the prize is discovery of a universe vastly greater than ourselves. By making ourselves part of that universe, we elevate ourselves to a new plane of conscious experience. "There must be new contact between men and earth," says the naturalist Wendell Berry; "the earth must be newly seen and heard and felt and smelled and tasted; there must be a renewal of the wisdom that comes with knowing clearly the pain and the pleasure and the risk and the responsibility of being alive in this world."

Order and Chaos

The binomials of Linnaeus were a liberating step that opened the path to the idea of evolution and common descent. But to live within the new evolutionary universe does not mean we must live in the stuffy world of Latin binomials. The art of reasoning may be nothing more than well-arranged language, but the lively chaos of common names is closer to the immediacy of daily life. The swallowtail may be *Papilio zelicaon* to the professional lepidopterist, but it will always be a swallowtail to the rest of us. A sweetbrier rose known as *Rosa eleganteria* truly doesn't smell as sweet. The two systems of biological naming—common and scientific—are complementary and satisfy different parts of the human agenda, perhaps even different halves of the human brain. Seeing the two lists of names side by side on the pages of Jacqueline Miller's book gives a powerful sense of the creative energy that flows back and forth between them. Call them poetry and science, call them art and logic; we are only partly ourselves without both. The mind is repelled by too much randomness and stifled by too much order. We are Dionysian and Apollonian. Creativity thrives on a balance of the fluid and the firm.

An illustrative exercise: Which of the following strings of letters do you find most interesting?

1. Aaa aaa aaa aaa aaa aaa aaa aaa aaa aaa aaa aaa aaa aaa aaa aaa.

2. One fish two fish red fish blue fish. Black fish blue fish old fish new fish.

3. How sweet the moonlight sleeps upon the bank! Here will we sit and let the sounds of music creep in our ears.

4. Rot a peck of pa's malt had Jhem or Shen brewed by arclight and rory end to the reggin-brow was to be seen ringsome on the aquaface.

5. Vfg w eklpsi muc dvpk dbjhq a v sm i yu ncq bfox w wgbm ifiai lvdymssa lsa s s aiuro y astwae-qyw rtwvme gv k ljr jxbkdq.

Number 1 is pure repetition, presumably boring. Number 5 is chaos (I programmed my computer to generate random letters and spaces); not much interesting there. Young children will prefer number 2, a passage from the Dr. Seuss book *One Fish Two Fish Red Fish Blue Fish* with lots of rhythm and simple pattern. A few adults profess to enjoy number 4, from James Joyce's *Finnegans Wake*, full of complex, deeply buried patterns. My guess is that most readers picked number 3, a snippet of Shakespeare. The human mind is most at home somewhere between perfect order and perfect chaos.

And thereby hangs a tale.

Linguists tell us that all human languages are about equally complex; the level of complexity is roughly equivalent to number 3. It would be reasonable to suppose that the complexity of languages matches the complexity of the human brain; that is, the richness of vocabularies and grammars is about that which the brain can process efficiently and without intolerable errors. Further, it would be reasonable to suppose that the complexity of the brain matches the average complexity of the human environment. Our brains evolved in a world that is balanced somewhere between perfect order and perfect chaos. Our neural systems were adapted by natural selection to recognize and process patterns in the sensed environment. If we are correct in these suppositions, then language sample number 3, from Shakespeare, is a pretty good match for nature's own balance of order and chaos.

Some parts of nature—the structure of crystals and the motions of the planets, for example—are close to number 1. Their patterns are massively repetitious and easily described. This is the physicist's domain, where science has had its most stunning successes. This is the domain of the strong reductionist, where mathematics has its broadest and most successful application. Certain other parts of nature—the weather and dreams, for example—are closer to number 5. Their patterns are hidden and complex, and have so far resisted complete analysis. Not pure chaos certainly, but as thickly contingent upon deep structure as *Finnegans Wake*. Which brings us to one of the most profound controversies raging in science today.

Some scientists—including "strong reductionists"—believe that nature's fundamental laws will be discovered near the number 1 end of the spectrum. As we have seen, Steven Weinberg, Stephen Hawking, and others hope to capture the totality of creation with a few simple mathematical equations, the Grand Unified Theory, the GUT of the universe. High-energy physicists and cosmologists like to imagine that in the first moments of the Big Bang the universe had a kind of perfect mathematical simplicity that quickly fractured as the universe cooled into a few fundamental forces that continue to govern everything that happens in the world today. In this strong reductionist view, life, intelligence, weather, and dreams will eventually be explained by physics, when we learn more about them.

Other scientists—call them "complexologists"—believe that the number 5 end of nature's spectrum has laws of its own, and no bottom-up explanations will suffice. This group is best exemplified by scientists at the Santa Fe Institute in New Mexico who are exploring the new mathematics of chaos and complexity. More is different, they say. When systems of interacting elements (particles, molecules, neurons, etc.) reach certain levels of complexity, phenomena emerge that cannot be predicted from the behaviors or laws of the separate parts. Weather and dreams, life and intelligence may be examples of nonreducible emergent phenomena.

Strong reductionist science owes much of its

success to the "well-arranged" languages that have been constructed at the simple end of the spectrum. For all of the hype coming out of Santa Fe, complexologists have so far added very little to our understanding of how the world works. But they are confident, and my guess is that the future of science will lie increasingly in their direction. Powerful, high-speed computers will be the key to the coming paradigm shift, providing a kind of "well-arranged" discourse appropriate to hugely complex systems. If the neural networks of the human brain have evolved to match the *average* level of complexity in our sensory environment, then deeper levels of complexity (consciousness, organic development, etc.) will be understood only by supplementing the brain with artificial intelligence. This is already happening. Computers are extending the powers of the human brain, transforming the way science is done. It seems inevitable that computers will give rise to wholly new ways of thinking about the world—new kinds of "well-arranged" top-down science.

It is not hard to imagine that someday we will have computers for which *Finnegans Wake* is a work of Dr. Seussian simplicity, and which equal in their own complexity currently intractable patterns of life and mind. At that point, we will have bootstrapped ourselves above the limitations of a brain adapted by evolution to a perceptual world of merely Shakespearean complexity.

The Struggle for Objectivity

Like most people who make a living communicating science, I spend long hours reading the scientific literature. In a typical week, I peruse several books and a dozen journals. It's not always fun. Scientific literature—scientists talking to scientists—is dull by design. It's like Sergeant Joe Friday said: "The facts, ma'am, just the facts."

For example, I've been reading a series of reports in the journal *Science* on the Galileo Probe, which was parachuted into Jupiter's atmosphere on December 7, 1995. Twenty-four pages of tough slogging, some of which passed right over my head. Tables of entry parameters. Doppler measurements of wind velocities. Pressures and temperatures. Mass spectrometer analysis of atmospheric gases. Refractive indexes. Solar and thermal radiation. Atmospheric scattering. High energy charged particles. Radio frequency signals. Numbers, graphs, tables, formulas, diagrams. A sample sentence (I will set it off in italics so it will be easy to skip):

> *A complete calculation of the He mole fraction q_{He} needs to take into account quantitatively (i) the pressures of the sample gas P_s and the reference gases P_r (or instead of the latter, the pressure difference between the sample and reference gases) at the start and end of the measurement in the jovian atmosphere; (ii) the*

absolute temperatures of the same gas T_s and the reference gas T_r at the start and end of the measurements; (iii) the Lorentz-Lorenz function that connects the refractive index n of a nonpolar gas with its mass density p(ρ); (iv) the non-ideal gas characteristics of H_2, He, Ar, and Ne as described by their compressibilities Z and virial coefficients B(T); and (v) the effects of an absorber in front of the SGC, which eliminates the traces of jovian methane from the measured gas sample.[1]

This dreary symbolic language has a purpose. Scientific literature emphasizes the part of our experience that is common to anyone who makes the observations in the same way. The quantitative prose of science is a way of separating the world "out there" from the world "in here." Ordinary language is laden with cultural and personal baggage. It transfers too much of ourselves onto the thing described. The struggle for objectivity is what makes science a source of reliable knowledge.

Yet objectivity is not enough to satisfy. We are creatures of the middle. Our minds have evolved in a world measured by the cadences and imagery of Shakespearean prose. We are emotional creatures. We have appetites. We are driven by awe, terror, love, hate. Our interior world is not always (ever?) "well-arranged." A diet of purely objective knowledge is oppressive. The poet Mary Oliver writes:

Still, what I want in my life
is to be willing

to be dazzled—
to cast aside the weight of facts

and maybe even
to float a little
above this difficult world.
I want to believe I am looking

into the white fire of a great mystery.[2]

Oliver does not denigrate facts. Her poetry is filled with precise observations of the natural world that match in their exactitude those of any scientist; this is one reason I find her work so attractive. Her exact knowledge of nature is the springboard from which she dives into the white fire of mystery. She asks to be willing to be dazzled. She asks for more than the weight of facts.

A white fire burns in those twenty-four fact-filled pages of data and analysis from the Galileo Probe—if we are willing to see it, or allow it to be interpreted for us. *A spacecraft, named for a hero of human intellectual freedom, is hurled from our planet on a six-year voyage across the empty darkness to Jupiter. It arcs toward Venus, rendezvousing briefly with that planet, then swings by Earth again, getting a boost from the slingshot of our planet's gravity. While passing by, it photographs the Earth and the Moon from space, two marbles afloat in India ink. Outward now, to an encounter with the asteroid Gaspra, sending back pictures of a city-sized potato pocked with craters. Back to Earth again, another gravitational boost. Then, on its definitive way at last, it passes the asteroid Ida, discovering that Ida has a tiny moon attached to the asteroid by a gossamer-thin gravitational thread. Approaching Jupiter from its unlit side, Galileo photographs the impact of Comet Shoemaker-Levy 9 as it smashes into the planet, an event hidden from observers on Earth. Finally, encountering Jupiter precisely as was planned six years earlier, on schedule, more than half a billion miles from Earth, Galileo goes into orbit around the giant planet. The craft releases a tiny human-built machine that parachutes into the jovian murk, sending back to Earth a stream of data—knowledge as esoteric as any cabala—as it plunges to oblivion.*

Facts, yes, a flood of facts. But more! For a moment, we are allowed to float above this difficult world, with Galileo's dazzling images of blue-white Earth suspended in darkness, pockmarked Gaspra, Ida with its miniature companion, and Jupiter, the giant planet, aswirl with color, a maelstrom and forge of creation, the object of human consciousness extended into cosmic space and time—the white, white fire of a great mystery, an intimation of the God of galaxies.

Knowing and Believing

Cosmology, spirituality, celebration—these are the attributes of religion. Each depends upon the others for the fullness of its expression. Cosmology reveals the creation; it answers the big questions: Who am I? Where did I come from? Where am I going? For better or worse, this is the task of science, exercising a way of knowing invented by the Greeks, perfected during the Scientific Revolution and Enlightenment, now embraced globally as the one truly human instrument of cosmic revelation.

For the method to work, we pretend for the moment that it is possible to step out of ourselves into the world *as it is*. To this end we invent names—*Cercyonis pegol*, *Cercyonis meadii*—that match the patterns we think we see in nature, or perhaps it is better to say that we choose names that are transparent to the patterns in nature. (The fact that Linnaean binomials are in Latin, a dead language, is relevant; this is one more way to step out of ourselves into an approximation of objectivity.) The language of science is transparent by design; wherever possible we engage the symbolic language of mathematics, doughty little *x*, the empty mirror, ready to reflect.

Of course, *perfect objectivity is impossible*. The poet Howard Nemerov says of poetry: "Poetry works on the very surface of the eye, that thin, unyielding wall of liquid between mind and world, where, somehow, mysteriously, the patterns formed by electrical storms assaulting the retina become things and the thoughts of things and the names of things and the relations supposed between things." Science works there too, in that wall of liquid between mind and world. With poetry, the arrow of transference is outward, from mind to world, the soul gone exploring in the white fire of mystery; with science, the arrow of transference is inward, from world to mind, a soul-making vector, incandescent with facts, sparks of the white fire kindled in our hearts.

If the prodigious energy of the new scientific story of creation is to flow into religion, the story will need to be translated from the language of scientific discovery into the language of celebration. This is the work of theologians, philosophers, homilists, liturgists, poets, artists, and, yes, science writers. Only when we are emotionally at home in the universe of the galaxies and the DNA will the new story invigorate our spiritual lives and be cause for authentic celebration. Knowing and believing will come together again at last. Cautious and skeptical as knowers, we can then give ourselves *unreservedly* to spiritual union with creation and communal celebration of its mysteries.

THIRTEEN

Work of the Eyes, Work of the Heart

To a poet, nothing can be useless. Whatever is beautiful, and whatever is dreadful, must be familiar to his imagination; he must be conversant with all that is awfully vast and elegantly little.

Samuel Johnson

IN THE SPRING OF 1996, we watched the long slide of Comet Hyakutake from the dark southeastern sky, up across the north pole, lapping Polaris with its streaming tail, into the light of the setting sun. It was a journey of exceptional duration, long enough for the comet to become a familiar part of our lives, the brightest comet I had seen since Comet West in 1976 and the brightest to grace the evening sky since Comets Arend-Roland and Mrkos in 1957. By the time it slipped into the twilight, Hyakutake seemed as if it had been with us forever. The comet was an easy naked-eye object. On the best night we saw the tail extending a full handspan across the sky—like the smudge of a fingertip on the windowpane of night—and this from the suburbs of the city.

On most clear nights, I was at the college observatory, together with a crowd of eager watchers drawn by the campus grapevine. Apparently, comets have not lost their old power to excite the imagination—to spook, to exhilarate. The turn-of-the-century naturalist John Burroughs wrote: "The night does not come with fruits and flowers and bread and meat; it comes with stars and stardust, with mystery and nirvana." The comet did not offer pyrotechnics, no billowing, sky-brightening special effects, just faint light and beauty. Most of my companions in the observatory were genuinely excited to see Hyakutake, and especially to stand in the cold dark with others who had come to celebrate nature's capacity to dazzle with stardust and nirvana.

It was, I submit, an experience as close to the primal origins of religious feeling as we can get in this increasingly virtual world. We watched in awe. We celebrated. We partook of mystery. No longer are comets portents, warnings, harbingers of doom.

Today we calculate their precise orbits—months, years, centuries in advance. We encounter them with spacecraft. We examine their substance with spectroscopes. We probe their gases. In all of this, the comet is a revelation of the mathematical order that underlies creation, of the material foundations of life and consciousness in the dark forge of space.

Comets are the stuff of which the solar system was born, visitors from the Oort cloud, that cold storehouse of icy snowballs believed to be orbiting the Sun in the realm beyond Pluto, containing perhaps as many as several billion comets with a total mass about that of the Earth. Hyakutake is an emissary from that unseen world of cold dark matter. Volatile compounds. Carbon-based molecules. Amino acids. The building blocks of life. Some scientists believe the materials of life were rained down upon the Earth by comets in the early eons of the solar system, 4 billion years ago. A few have suggested that life might have come to our planet on a comet as primitive microorganisms. Watching the faint light of Hyakutake, we were witness to the kindling of life in the dark hearth of space, the faint glow of a flame that in human consciousness flares to illuminate the galaxies.

I first saw Hyakutake from an island in the Bahamas in mid-March. I wasn't looking for it; I hadn't expected the comet to become a naked-eye object until later on, when it would be higher in the east.

But there it was, an easy naked-eye object in Virgo. At first I thought I might be looking at M5, a globular cluster of stars purportedly visible to the unaided eye under perfect conditions, and a permanent resident of that part of the sky. But the next night, when the blur had moved, it was obvious that Hyakutake was on its way and that it was going to be an exceptional apparition. Night by night the comet grew brighter, flirting with Arcturus, skimming the Dipper. We watched it with binoculars and telescope, but the unaided eye was the perfect instrument, allowing the comet its majestic context, a far-flung tail, a backdrop of stars, an abyss of darkness.

Best of all was the evening of April 3, when we forsook the observatory for a broad dark field where we watched the Moon rise in full eclipse, a spooky pink pearl. The comet was in the northwest, showing a degree or two of tail. Venus had joined the Pleiades, a blazing beacon. Meteors streaked the firmament. I was with a group of young people, students at the college. I was impressed by their reverence, wonder, worship even—and especially by their intense desire to *know*. I described the physics of cometary motion and produced a three-dimensional model of the orbit I had previously constructed. We talked about the chemistry of comets and the chemistry of life. Knowledge, wonder, and celebration played off each other in perfect harmony. I thought: How sad that such experiences are not part of our formal religious traditions. It was at that moment, in that field, watching that comet, that I decided to write this book.

We are a culture of special effects, virtual reality, ersatz experience. Generally, it takes a blockbuster to gain our attention. Super Bowls. Mega-events. Las Vegas and Orlando. To bring us out at night away from our big-screen TVs, one would have supposed the sky would have to roil with coruscating light. Instead, the comet whispered. And a crowd of young people was there to listen. Hyakutake spoke sweet nothings; they cocked their ears. A smudge on the windowpane of night; they nodded approvingly. With sublime discretion, the comet crossed the black deep, trailing a wake of exquisite fineness, and the audience was sitting on the edge of their seats.

So maybe we are not yet immune to the primordial epiphanies that ignited the spark of worship in the minds of our ancestors. John Burroughs said that "the good observer of nature exists in fragments, a trait here and a trait there." And again, "One secret of success in observing nature is capacity to take a hint." Watching Hyakutake, we waited for traits and took the hints. Knowledge did not diminish the experience. Knowledge and wonder and celebration went hand in hand in hand: the archetypal religious experience.

The Roman Catholic priest and cultural historian Thomas Berry, who urges upon us what he calls the New Story, the story of science, writes: "Future and past cannot live off the present forms of religious experience for these are too shallow; the future can live only from the most primordial communion with the sacred."[1] We stood in the dark field, the moon in eclipse, Venus in the Pleiades, the comet arching over all, and a shiver went up our spines. We were engaged in collective worship, in communion with the God of galaxies. Our minds were mirrors of divinity. When night's faint lights revealed themselves to Burroughs, his thoughts went "like a lightning flash" into the abyss, and then the veil was drawn again. It was just as well, he wrote, to have such faint and fleeting revelations of the deep night: "To have it ever present with one in all its naked grandeur would perhaps be more than we could bear."

A Snowstorm of Galaxies

Hyakutake was not the only spectacular celestial revelation of the spring of 1996. There was another historic revelation, less immediate than the comet but no less grand: the Hubble Space Telescope Deep Field photograph of the early universe. Many newspapers and magazines published a version of this extraordinary image of the most distant galaxies ever observed, but these were mostly black-and-white partial reproductions that did little justice to the real thing. To view the image to full effect, one should project the color slide onto a big screen in a pitch-black room. Then, it is like looking into the soul of the night.

The Hubble Space Telescope focused its camera on a tiny speck of sky, chosen more or less at random from the apparently empty space between the visible stars, for an unprecedented ten days, through 342 separate exposures, soaking up the faint light of galaxies beyond the range of Earth-based telescopes. The result: a breathtaking snowstorm of galaxies, in living color, including a few relatively nearby spirals and hundreds of faraway galaxies that show up as mere specks of light. Because light takes time to reach us, when we look into deep space we are looking back in time. The most distant galaxies in the photograph are more than 10 billion light-years away. We see them as they were not long after the universe's beginning.

Go out tonight under the starry sky with a common pin in each hand and cross the pins at arm's length. The intersection of the pins against the background sky is the area shown in the Hubble Deep Field photograph. It would take 25,000 photographs at this scale to cover the bowl of the Big Dipper. To make the image, the Hubble Space Telescope was pointed to a part of the sky that reveals absolutely nothing to the naked eye or even to a small telescope. The field is typical of what we would see if we looked in any direction into the universe. In that tiny area of the sky—the intersection of two pins held at arm's length—the photograph shows at least 1,500 galaxies. A survey of the Big Dipper's bowl at the same level of detail would show nearly 40 million galaxies, and a survey of the entire sky would reveal 50 billion galaxies. That's as many galaxies as there are grains in several thousand one-pound boxes of salt.

Each galaxy contains hundreds of billions of stars. Many of those stars have families of planets. Our Sun is just one star in the Milky Way Galaxy, a spiral of a trillion stars. Think of the Milky Way Galaxy as a dinner plate. The next spiral galaxy—the Great Andromeda Galaxy—is another dinner plate across the room. At this same scale, the nearest galaxies in the Hubble Deep Field photograph are dinner plates about a mile away. The faintest are dinner plates more than twenty miles away, at the very edge of space and time.

What is the *scientific* significance of the new Hubble photograph? As I write, it is too early to tell. Astronomers hope that by seeing deeper into space they will learn more about the earliest days of the universe, including the origin and evolution of the galaxies. They also hope to learn more about the deep structure of space and time. That's a lot to ask of a single photograph of a tiny speck of sky, but it's a start. For the moment, however, the Hubble Deep Field photograph is primarily a religious icon that expands our horizons and sharpens our sense of the size and prodigiousness of the universe.

The new view of deep space shows five times more galaxies than had been seen in such a sample before, taking us closer to the presumed epoch when the galaxies condensed from the primordial fire. Take a look at the Hubble Deep Field photo and then go outside and hold those crossed pins against the night sky. Let your imagination drift away from the Earth into those yawning depths where galaxies whirl like snowflakes in a storm. From somewhere out there among the myriad galaxies, imagine looking back to the one dancing flake that is the trillion stars of our Milky Way. Galaxies as numerous as snowflakes in a storm!

When the Hubble Deep Field photo appeared in the newspapers, some people came up to me and said, "It makes me feel so insignificant." No, no, I insisted, that's exactly the opposite of what we should feel. The Hubble Deep Field photo is a product of human science, the culmination of thousands of years of wondering at the night sky. It is the immediate creation of millions of persons of many nationalities, including the research scientists who analyze the data, the machinists who fashioned the nuts and bolts, and folks like you and me who support the instrument with our tax dollars. We believe that what the photograph reveals is actually *out there* pretty much as we see it in the photo. But a transfer has been made across that thin film in the eye that separates the mind from the world. The universe of the photograph also exists *in here*, in our minds. We carry 50 billion galaxies in our heads, and that makes us pretty significant, it seems to me.

This new knowledge of the galaxies is exhilarating and terrifying, beautiful and dreadful. Yes, it can be emotionally deflating knowledge if we continue to insist that *we* are the measure of all things. But if we accept that the human physical scale is an inadequate measure for creation, then the Hubble Deep Field photo opens us to a cosmos of capacious grandeur. A universe of 50 billion galaxies blowing like snowflakes in a cosmic storm is astonishing, but even more astonishing are those few pounds of meat—our brains—that are able to construct such a universe of faint light and hold it before the mind's eye, live in it, revel in it, praise it, wonder what it means. The fourteenth-century mystic Julian of Norwich asked, "What is the use of praying if God does not answer?" In that wonderful image of more than 1,000 galaxies, caught by a magnificent instrument lofted into space by a questioning creature, God answers.

To Love What Is Mortal

The Copernican and Darwinian revolutions were liberating breakthroughs. The first removed us from the center of space, the second from human time. These twin foundations of the New Story have brushed away the last cobwebs of animism, anthropomorphism, anthropocentrism. The human gods are swept from their thrones. Angels, devils, spirits, and shades are sent packing. We are contingent, ephemeral—animated stardust cast up on a random shore, a brief incandescence.

As we have seen, the majority response to the New Story is denial. We are grudgingly willing, many of us, to entertain the notion of the galaxies and the galactic eons in the backs of our minds, in a closed cupboard called "science." But we vehemently resist the human implications of the New Story. We plead "faith," "revelation," and "tradition" as the bases for rejecting the New Story, forgetting that faith, revelation, and tradition, if they are to mean anything at all, must be consistent with what our senses tell us is true. The reason for our resistance is plain, indeed well-nigh irresistible: death. The New Story makes one thing clear: *We are not immortal.* Our selves are fleeting. Our substance flows through the body of the world. Our spirits are the brief efflorescence of complexity. Our response to this new knowledge must be what the poet Mary Oliver suggests in a poem called "In Blackwater Woods":

> *. . . To live in this world*
>
> *you must be able*
> *to do three things:*
> *to love what is mortal;*
> *to hold it*
>
> *against your bones knowing*
> *your own life depends on it;*
> *and, when the time comes to let it go,*
> *to let it go.*[2]

If we can surrender the ancient dream of immortality, then we can begin building a new theology, ecumenical, ecological, non-idolatrous. It will emphasize our relatedness and our interrelatedness, our stewardship rather than our dominion. It will define our value by our participation in a cosmic unfolding; we are flickers of a universal flame—galaxies, stars, planets, life, mind—a seething cauldron of creation. Natural and supernatural, immanent and transcendent, body and spirit will fuse in one God, revealed in his creation. We have discovered the story on our own. On this speck of cosmic dust, planet Earth, the universe has become conscious of itself. The creation acknowledges the Creator. Our lives are sacramental. We experience the creation in its most fully known dimension. We celebrate. We worship.

In a poem titled "He Wishes for the Cloths of Heaven," William Butler Yeats muses:

> *Had I the heavens' embroidered cloths,*
> *Enwrought with golden and silver light,*
> *The blue and the dim and the dark cloths*
> *Of night and light and the half-light,*
> *I would spread the cloths under your feet.*[3]

Few more beautiful words have been put on paper than these lines of Yeats's. But we need not merely wish for the cloths of heaven. They are ours, now. The gorgeous blue-dim tapestry of night is spread from horizon to horizon, studded with diamond lights and embroidered with the golden and silver threads of the Milky Way. The cloths of night have inspired religion, myth, mathematics, and science since the first spark of consciousness ignited the human brain.

Even today, in our technically sophisticated times, a view of the night sky from a dark place—Hyakutake on its westward arch, Venus among the Pleiades, the Moon rising in eclipse—cannot fail to inspire dreams of a grandeur and a meaning greater than ourselves. But there is more, much more. Through our science we have created magnificent spacecrafts and telescopes to explore the night and the light and the half-light. We have made visible things that are invisible to the unaided eye. We have brought the dreamy heavens down to Earth, held them in the mind's eye.

Our explorations have produced a vast archive of remarkable astronomical images. Among my favorites:

- The Earth and Moon in the same field of view, imaged by the Galileo spacecraft on its second flyby of Earth. The Earth is delicately gauzed with water and air, the Moon dull and lifeless.

- The pocked, potato-shaped asteroid Ida and its baby moon Dactyl, imaged by Galileo on its way to Jupiter. Ida is a bit smaller than Rhode Island; Dactyl could rest comfortably in New York City's Central Park.

- The Voyager photographs of Jupiter and its four largest moons. Nothing could have prepared the imagination for the colors and textures of these psychedelic worlds.

- The Hubble Space Telescope image of the Cygnus Loop, wispy shreds of a star that blew itself to smithereens 15,000 years ago, a lacelike delicacy that belies the violence of its creation.

- The Hubble image of a gassy star-factory in Orion, a myriad of worlds coming into existence before our very eyes.

- The head-on collision of a small galaxy with a giant spiral galaxy in the constellation Sculptor, imaged by Hubble. Like a stone thrown into a pond, the collision hurls outward a shock wave that ignites a brilliant tiara of stars.

- Light of faraway galaxies curved into a concentric swarm of circular arcs by the gravity of a massive, compact cluster of nearby galaxies, a rose-window view of the universe's youth and a mind-blowing confirmation of Einstein's theory of general relativity.

- And, of course, the Hubble Deep Field photograph.

But why am I choosing favorites? The riches are too many for choices, the revelations beautiful and dreadful. Who can look at these images and not be transformed? The heavens declare God's glory. "The work of the eyes is done, now / go and do heartwork," says the poet Rainer Maria Rilke.

NASA should produce a full-length film of astronomical images for the big screen, a grand tour of the universe, from Earth to the black holes at the cores of distant galaxies, a Hollywood-quality production with narration by Leonard Nimoy and score by John Williams. Cinemas in every town and neighborhood of the world should be subsidized by governments to show the film for free. Yes, it would cost several hundred million dollars, but that's a small fraction of the cost of the scientific programs that produced the images. These splendid products of human curiosity and ingenuity could be the true Gothic cathedrals of our time, the nexus where human striving touches the highest mysteries—the embroidered cloths of heaven laid at the feet of every human, rich or poor.

Knowing and Loving

As we stood in the dark field watching Comet Hyakutake, we were not the first who had watched a comet. But we were different. We were post-Copernican, for one thing, and post-Darwinian. We knew the comet was not meant for us. It was not a portent. It followed a preordained ellipse—mathematical, Newtonian. I said to my young companions, "The moon will rise just there, to the left of the tallest tree, already in eclipse." And it did, a thing of exquisite beauty, the color of mother-of-pearl. The difference was *knowledge*. And because our knowledge was different, our experience was different, our celebration was different.

We are not playthings of the gods, comet-warned and fearful; we are the comet's offspring, volatile compounds made animate, made conscious. We watched the Moon rise in Earth's shadow, Venus in the Pleiades, Hyakutake dipping westward. The universe took form in our minds, a snowstorm of galaxies. Our knowledge is immortal, a growing thing, the mark of our divinity—in our knowledge we will live forever. In the silence of that April night we heard what Christ said to Julian of Norwich: *This I am. I am what you love. I am what you enjoy. I am what you serve. I am what you long for. I am what you desire. I am what you intend. I am all that is.*

FOURTEEN

The Hallowing of the Everyday

The world is charged with the grandeur of God. It will flame out, like shining from shook foil.

Gerard Manley Hopkins

PINNED TO THE wall above my desk is a quote I found many years ago. I can't remember where I found it, or anything about the author except that his/her name is Dixon: "If there be a skeptical star I was born under it, yet I have lived all my days in complete astonishment." Science is founded on the twin cornerstones of skepticism and astonishment. Skepticism is a critical reluctance to take anything as absolute truth, even one's own most cherished beliefs. Astonishment is the ability to be dazzled by the commonplace. At first glance these two qualities might seem opposed. The Skeptic is often thought to lack passionate commitment. The easily astonished person is sometimes thought of as gullible. In fact, reasoned skepticism does not preclude passionate belief, and astonishment is enhanced by knowledge.

"Nothing is too wonderful to be true," said the nineteenth-century physicist Michael Faraday. And that is astonishment. But everything wonderful need not be true, and that is skepticism. The thoughtful person will try to walk the line between drop-jawed amazement at the wonder of creation, and cautious skepticism about the correctness or finality of our knowledge.

One morning not long ago, I walked to college through meadows made misty by the heat of the rising sun. As I rounded a stand of trees and stepped onto the footbridge over Queset Brook, I startled a great blue heron that stood not ten feet away. The heron startled me. It heaved into the air with bedsheet wings—*push, push*—I could feel the whoosh of air. Neck crooked, pennant head-feathers flying, legs dangling behind like loosened mooring lines. The size of it—our biggest bird! The fierce eye. The pterodactylian beak. The effect was prehistoric, like a scene from a movie, *Dinosaur Island* or *Jurassic Park.* I stood on the footbridge and applauded.

I'm no ornithologist, but I know certain things about herons that anyone might know, things accumulated by generations of ornithologists working patiently in the field, and by zoologists, anatomists, paleontologists, DNA experts, and aeronautical engineers. I know things that have been compiled in popular books by nature writers and field-guide authors. I know, for example, about the bird's feeding and mating habits, its voice and call, its relationship to the European heron and the Japanese crane. I know that the heron, like all birds, is a close relative of dinosaurs, and that feathered birds first flapped their wings in Jurassic times. There is nothing esoteric about any of this knowledge, nothing that requires special training in science. It can be found in sources like the *National Geographic* magazine, the Audubon Society's magazine, the science pages of the newspaper, or television nature shows. In the best of all worlds, it would be taught in the schools from an early age. It would be part of every child's intellectual inheritance, like nursery rhymes and fairy tales. Reliable knowledge, public knowledge, knowledge that enhances experience and increases wonder.

Everything I knew about herons was subsumed in that epiphanic moment when the bird lifted into the air, trailing its toes in the black water of the brook. The heron was feathered by knowledge. Its six-foot wings spanned continents; their beats marked eons of geologic time. In every cell of the bird's great body, coiled strands of DNA performed a dervish dance that can only be imagined in the mind's eye—spinning, unraveling, copying themselves—the kinetic miracle of life. In earlier times, myths and totemic religion would have provided the bird with a context, a human meaning. But the ancient myths and totemic religions no longer command our belief. Today, only scientific knowledge can weave the heron into a tapestry of larger meaning. For better or worse, science is the defining public knowledge of our time.

And what knowledge it is! A story of sublime dimension. Tentative, evolving, and not always comfortable, carrying us in our imaginations to the farthest reaches of space and time, but hedged about with death and oblivion. Scientific knowledge enlivens our every experience and tunes us in to the deepest mysteries of creation, the hidden rhythms of a world that evades our limited senses. Science cannot nor should not be a religion, but it can be the basis for the religious experience: astonishment, experiential union, adoration, praise. And so it was with the heron. As the dinosaurian relict pounded the air, I stood on the footbridge and gaped, seeing deeply into a world not altogether my own, totally skeptical, completely astonished.

Primordial Experience

I experience the heron with more than the eye and mind of science. I enter into a kind of communion with the heron, and feel—how can I say it?—a *mutuality* of relation. The fullness of the experience involves all that I am, all that I know, all that the heron is. Knowledge binds us—knower and known—yet the power of the experience is the sudden awareness of what is not known. I stand on the shore of knowledge and look to the far horizon of mystery. The mystery strikes deep. A shiver up the spine. Exhilaration and fear.

One of the formative books in my youth was Martin Buber's poetic and contemplative *I and Thou.* Buber was an eminent Jewish philosopher and religious scholar, born in Vienna in 1878. His most famous work was first published in English in the 1930s, but a second English edition was brought forward by Scribner in 1958, just at the time when I had begun my struggle to reconcile my scientific training with my childhood faith. What Buber offered (to me and others like me) was a vocabulary for understanding what we *felt*, a naming of two kinds of experiences, what Buber calls the *I-It* and the *I-Thou.*

Ordinary day-to-day experience—the experience of the scientific experiment, for example, or the experience I feel just now as I stare at the screen of my word processor—belongs to the realm of *I-It.* Such experience is necessary for living in the world; it is the basis for the practical agenda of living. We win our bread in the realm of the *I-It.* We put on our shoes, go to the bank, and change the oil in the car in the realm of the *I-It*. "Without *It* man cannot live," says Buber. And he adds: "But he who lives with *It* alone is not a man."[1] The kind of experience that is relational, mutual, and transcending can be classified as the *I-Thou*. My relation with the heron, for example. I see the heron as a sheet of feathers in a shock of light, a splash of blue shot through with silver, streaming droplets of gold. I perceive it as movement: wings heaving against air. I classify it as a species and study it as a type in its anatomy and mode of life. I subsume its presence into physics, chemistry, and molecular biology. I embroider the bird in pure mathematical relation. In all of this the heron remains an object, occupies space and time, and has its nature and constitution. (I paraphrase Buber.) In all of this the heron is an *It*. But it can also happen, *if I have both will and grace*, that in experiencing the heron I become bound up in an unbidden relationship. I am struck through by a power that resides in the bird, that finds a resonance within me—a power that is nameless, all-inclusive. I address the bird as *Thou*. I enter briefly, ecstatically into spiritual union with the bird, a union in which we are both subsumed into the greater mystery. The heron is no longer *It*. It has been seized by the power of exclusiveness. This relation—unasked for, unexplained—is the primordial religious experience.

There is a hierarchy of experiential relation

within the *I-Thou*: stone, plant, animal, human being—a spectrum of relational possibility. All of these lines of relation culminate in what Buber calls the eternal *Thou*. Every particular *Thou* is a glimpse through to the eternal. By means of every particular *Thou*, we address the eternal *Thou*, by other persons called God. But the word *God* has been so abused with idolatrous meanings that it can be helpful—for clearing the brush, so to speak—to temporarily substitute Buber's more intimate *Thou* that is and contains all things. And now Buber makes a statement of fundamental importance: "Men do not find God if they stay in the world. They do not find Him if they leave the world. He who goes out with his whole being to meet his Thou and carries to it all being that is in the world, finds Him who cannot be sought."

What is most illuminating about Buber's understanding of the *I-Thou* relationship is this: To experience the *I-Thou* with, say, the heron, it is not necessary for me to give up any of the ways in which I consider the heron. There is nothing from which I must avert my eyes, no knowledge that I must forsake. Trees, mist, rising sun, plank bridge—they all participate in the experience. The streaming feathers and dangling legs, the fierce reptilian eye, the beak—everything is indivisibly united in the *I-Thou*: outline and movement, species and type, law and number. Everything belonging to the heron is there: its form and structure, its colors and chemical composition, its intercourse with the elements and with the stars, are all present in a single whole. This is no play of imagination, no trick of mood. The heron has to do with me as I with it. In the *I-Thou,* we go beyond science, to something *felt* that cannot be expressed. I don't mean to evoke the supernatural or the transcendent. It is the strength of Buber's analysis that he emphasizes the *everyday* nature of the *I-Thou* experience. He writes: "The clear and firm structure of the *I-Thou* relationship, familiar to everyone with a candid heart and the courage to pledge it, has not a mystical nature. From time to time we must come out of our habits of thought to understand it; but we do not have to leave the primal norms which determine human thinking about reality."

I don't know how far to follow Buber in this: Although he explicitly seeks to avoid the pitfalls of animism, he teeters close to it with his notion of reciprocity in the *I-Thou* relationship with nonhuman nature. But on this much at least we can agree. The experience of the *I-Thou* is real. The perception of mutuality is real. The spoken *I-Thou* is real and felt to be different from the spoken *I-It.* We are changed utterly by the *I-Thou,* and the *felt* experience is that we are changed *by* the object of our address, struck, rung like a bell. The *I-Thou* takes us beyond *public* knowledge, beyond science, but not beyond knowing. We do not put a name to it. Any name is idolatrous; even, perhaps especially, given its history of casual abuse, the name of *God.* To experience the *I-Thou* we must be skeptical but open, knowledgeable but ignorant, waiting to be struck dumb, waiting for the grace of insight. "Creation happens to us," says Buber, "burns itself into us, recasts us in burning"—we tremble and are faint, we submit. We take part in creation, meet the Creator, reach out to Him as helpers and companions.

Why then do we hanker for miracles, aliens, UFOs, bleeding statues, angels, apocalypse? Why are we so quick to hand over our reason to the channelers, gurus, astrologers, faith healers, and promisers of a quick immortal fix? Because we have lost our sense of a community story, a story of who we are, where we came from, where we are going. The old story still holds our attention, but it is a hollow shell. The New Story waits in the wings. The scientists who have discovered the New Story have done a poor job of telling it, but perhaps it is not their task to be storytellers. The skills required for scientific discovery are not the skills of narration. In any case, the rest of us have not made much of an effort to listen, perhaps because we do not want to hear intimations of our own mortality. Lacking a community story that suffuses the world with unity and meaning, we turn away from the world *as it reveals itself*, and look to pseudoworlds and superstition. And in doing so, we lose access to the *Thou* that resides in *things*, an awareness of the primordial sacredness of creation. We have forgotten what Buber calls "the hallowing of the everyday."

"Men do not find God if they stay in the world," says Buber. Knowledge alone will not evoke him. We must leave the high ground of knowledge and walk the shore where knowledge is lapped by mystery. But neither will we find God if we leave the world, as do those who imagine that mystery alone will suffice. Abandoning the secure shore of knowledge, they swim into the sea, hoping to be raptured by aliens or angels or by a God who lives outside of his creation; they find nothing but the sea that fills their eyes and lungs. Buber writes: "To look away from the world, or to stare at it, does not help a man to reach God; but he who sees the world in Him stands in his presence. 'Here world, there God' is the language of *It*; 'God in the world' is another language of *It*; but to eliminate or leave behind nothing at all, to include the whole world in the *Thou*, to give the world its due and its truth, to include nothing beside God but everything in him—this is full and complete relation."

Creation Spirituality

Jesuit scientist Pierre Teilhard de Chardin was one of those rare individuals who experienced no tensions in his own life between knowing (science) and believing (theology). A field paleontologist, Teilhard traveled the world to search out fossil remains of human ancestors. For him, every fragment of fossilized bone and every flake of worked stone excavated from the earth was a clue to God's plan of creation. His vision of creation was evolutionary to the core: a preordained unfolding of life and mind from primordial matter. He looked for the completion of evolution in a kind of cosmic consciousness—the Omega, he called it—that he identified with God. These aspects of Teilhard's work did not always find favor among his scientific colleagues, who considered his speculations insufficiently empirical. Nor were they approved by Teilhard's ecclesiastical superiors in Rome, who found them not sufficiently transcendent and dangerously tainted with evolutionary science. Permission to publish his philosophical works was denied by Church authorities. He died in 1955 in New York City, exiled from his beloved France, swallowed up in a gaping chasm between his evolutionary science and the theology of his church.

Teilhard's best-known book, *The Phenomenon of Man*,[2] appeared posthumously in English in 1959, at a time of spiritual crisis in my own life. I was a graduate student in physics at the University of California at Los Angeles, seeking to reconcile the theology I had learned as an undergraduate at Notre Dame University with the science I was learning in one of the great secular institutions of learning. The two did not rest easily together. Teilhard blew into my life like a breath of fresh air. Here was a man who grounded his every thought and feeling in the facts of science yet professed a deeply religious view of the universe. The very first words of Teilhard's book swept me into their thrall: "To push everything back into the past is equivalent to reducing it to its simplest elements. Traced as far as possible in the direction of their origins, the last fibers of the human aggregate are lost to view and are merged in our eyes with the very stuff of the universe." The *stuff* of the universe! But not the inert atoms of the philosophers. Rather, Teilhard's "stuff" was charged with emergent possibilities, carried along by the swelling river of evolution from the Alpha of creation to the Omega of redemption. The human earthly drama was but a part of a greater story of creation and redemption that embraced the galaxies, arched the eons. Theology must not merely *accommodate* evolution, he believed; it must take evolution as its starting point. His vision convinced me that religion and science might be reconciled after all.

Recently, I returned to Teilhard's *The Phenomenon of Man* for another look. It is easy to see now what exasperated his scientific colleagues. "If this book is to be properly understood, it must be read . . . as a scientific treatise," Teilhard writes in the preface. But in fact the book is a tissue of nonempirical intuitions, couched in words—*noosphere, the cosmic law of complexity-consciousness, the Omega*—with only the vaguest scientific usefulness. *The Phenomenon of Man* is far too mystical to appeal to scientists, and too worldly to satisfy traditional theologians. It is the intensely personal confession of faith of a God-struck dreamer, hopeless as a program for bridging the gulf between science and theology. Yet many of Teilhard's intuitions strike me as on the mark. The chasm between theology and science remains wide and deep, as needful as ever of being bridged. Our imaginations remain by and large transfixed by the same ancient distinctions between matter and spirit, natural and supernatural, profane and sacred that Teilhard rejected.

In this, at least, Teilhard cannot be faulted: He insisted that the surest way to know God is through his creation, and the truest knowledge of creation is that provided by contemporary science. Late in life he wrote to a friend: "Less and less do I see any difference now between research and adoration."

Like Shining from Shook Foil

When I was a very young child, a picture of an angel hung on the wall above my bed, a beautiful winged creature guiding a boy and girl across a rickety footbridge. It was, of course, a guardian angel, and according to my parents each of us has one. Before I went to sleep I said the traditional prayer that begins "Angel of God, my guardian dear . . ." It is a consoling idea, that one of the heavenly choir is assigned to each of us, to guide us safely across the rickety bridges of life and watch over us as we sleep. My own guardian angel hovered reassuringly at my side until about the time I went off to school, slipped from my consciousness at adolescence, and vanished completely as I began the study of science.

Angels and skepticism don't mix. But angels are hot properties at the end of the millennium. Hollywood has given us a slew of angel movies. Angelology has moved from theological texts to the supermarket newspapers. Bookstores sell angel books by piles. Suddenly, it seems, angels are everywhere. You can almost hear the din of their fluttering wings. This premillennial proliferation of supernatural spirits is symptomatic of our civilization's rift between knowing and believing.

Our technology, economic well-being, long lives, and generally good health are based upon a way of knowing that values skepticism and the evidence of the senses. One could argue that our political freedoms, so profoundly influenced by the Enlightenment, are grounded in a skeptical view of the world. However, a growing number of us are suspicious of science, nostalgic for a world animated by spirits, and given to the notion that each of us has a direct personal line to whatever power rules the universe. We accept science for the material benefits it contrives on our behalf, but we distrust the materialist/naturalist philosophy on which science is based, preferring to give our attention to anyone claiming commerce with angels or aliens. We turn to science to remedy our ills but are quick to blame science for our misfortunes. In our schools we teach kids the public knowledge of astronomy, biology, chemistry, and physics, and in our homes we follow private visions of astrology, creationism, health fads, and parapsychology. We live *in the world* six days of the week and *out of the world* on Sundays. In a word, we are riven.

If a medieval philosopher were confronted, on the one hand, with the idea of angels and, on the other, with the idea of air resonant with 100 species of unheard music (Bach, Beethoven, and Mozart, to say nothing of Frank Sinatra and the Grateful Dead) made audible by a small box called a radio, he would surely call the latter the more miraculous. And, I suspect, he would find the idea of our humanness revealed by molecular biology and neurobiology more godly and inspiring than John Donne's "little world made cunningly of Elements, and an Angelike spright." If we abandon ourselves to True Belief, we revert to a world animated by angels and devils, portents and omens, possessions, flagellations, inquisitions. At risk are the material and social fruits of the Scientific Revolution and the Enlightenment; waiting in the wings are the infantile forces of sectarian strife and superstition. If history is a guide, the split between Skeptics and True Believers will grow wider as we approach the end of the millennium. The angels invading our bookstores and cinemas announce a binge of otherworldliness on the countdown to the apocalypse. Science is merely tolerated; pseudoscience and fundamentalist religions are passionately embraced. And those of us who have chosen to negotiate the rickety footbridge of life with only a heron at our side will find ourselves increasingly alone.

But perhaps I am reading history too pessimistically. We are a creative, hopeful, resourceful species. We have renewed ourselves before. The pieces are in place for a renaissance of religion: cosmic knowledge, the power for good, awareness of mystery, a sense of responsibility to all of creation, and a longing for union with the Absolute. What is required is imagination, self-confidence, courage. The world is charged with the grandeur of God. If we are worthy, it will flare out, in the words of the poet Gerard Manley Hopkins—like shining from shook foil.

Notes

Introduction

1. John Polkinghorne, *The Faith of a Physicist* (Princeton, N.J.: Princeton University Press, 1994), pp. 14 and 5.

1. Miracles and Explanations

1. Gerard Manley Hopkins, "The Starlight Night," *The Poetical Works of Gerard Manley Hopkins*, ed. by Norman H. MacKensie (Oxford: Clarendon Press, 1990), p. 139.
2. Richard Dawkins, "Putting Away Childish Things," *Skeptical Inquirer*, January–February 1995, p. 139.
3. Living organisms build their bodies by taking carbon from the atmosphere. There are two kinds of carbon atoms, called isotopes. Carbon-12 has six protons and six neutrons in its nucleus; carbon-14 has two extra neutrons. Carbon-12 nuclei are stable; carbon-14 nuclei are unstable, or radioactive, and disintegrate at a precisely known rate called a half-life (the half-life of carbon-14 is 5,568 years). In spite of the decay, the ratio of carbon-14 to carbon-12 in the atmosphere is approximately constant, maintained by the creation of radioactive carbon-14 nuclei by cosmic rays from outer space. When an organism dies, the carbon-14 in its structure decays without being replaced, changing the ratio of isotopes in a clocklike fashion. The number of

atoms of both species in a sample can be determined using a mass spectrometer.

4. P. E. Damon, D. J. Donahue, and B. H. Gore, "Radioactive Dating of the Shroud of Turin," *Nature* 337 (February 16, 1989): 311–15.
5. Pope John Paul II, "Science and Faith," address at the University of Pisa, September 24, 1989. Reprinted in *Origins, Catholic News Service Documentary Service* 19 (October 26, 1989): 21.
6. John Donne, Sermon, Easter Day, March 25, 1627, *The Complete Poetry and Selected Prose of John Donne*, ed. by Charles M. Coffin (New York: Modern Library, 1952), p. 536.
7. For the story of the red knot, I am indebted to Brian Harrington, *The Flight of the Red Knot* (New York: W. W. Norton, 1996).

2. Decoding the Mystery of Life

1. A photograph of tangled DNA can be found in *Science* 247 (February 23, 1990): 913.
2. Henry Margenau, *The Nature of Physical Reality* (New York: McGraw-Hill, 1950).
3. *Science*, 270 (December 8, 1995).
4. William Faulkner, Nobel Prize acceptance speech, Stockholm, December 10, 1950.
5. Gerard Manley Hopkins, "The Starlight Night," *The Poetical Works of Gerard Manley Hopkins.*

3. The Known and the Unknowable

1. Chet Raymo, *Honey From Stone: A Naturalist's Search for God* (St. Paul, Minn.: Hungry Mind Press, 1997), p. 58.
2. Erwin Chargaff, "Engineering a Molecular Nightmare," *Nature* 327 (May 21, 1987): 199.
3. Richard Feynman, *No Ordinary Genius*, ed. Christopher Sykes (New York: W. W. Norton, 1994), p. 107.

4. Organized Skepticism

1. Nicholas Humphrey, *Leaps of Faith: Science, Miracles, and the*

Search for Supernatural Consolation (New York: Basic Books, 1996), p. 10.

2. Anthony Storr, *Feet of Clay: Saints, Sinners, and Madmen: A Study of Gurus* (New York: Free Press, 1997), p. 203.
3. Blaise Pascal, *Penseés*, 68 (206).
4. Alan Lightman and Owen Gingerich, "When Do Anomalies Begin?" *Science* 255 (February 7, 1992): 690.
5. For comment of firewalking enthusiast, see Michael Sky, *Dancing with the Fire*, as excerpted on the Heartfire Communications website, 1995.
6. John Henry Newman, *The Idea of a University* (New York: American Press, 1941), p. 445.

5. Astrology and Prayer

1. For measures of scientific literacy in America, see the 1989 survey by Northern Illinois University Public Opinion Laboratory, reported in *Science* 243 (February 3, 1989): 600. See also *Science* 251 (March 1, 1991): 1024, and the 1994 survey by the American Museum of Natural History and Louis Harris & Associates, reported in *New York Times*, April 21, 1994, p. D23.
2. For discussion of controlled tests of astrology, see Geoffrey Dean, "Does Astrology Need to Be True?" *Skeptical Inquirer* Winter 86/87, p. 166, and Spring 87, p. 257.
3. *Time*, June 24, 1996, p. 40.
4. Randolph Byrd, *Southern Medical Journal* 81, no. 7 (July 1988): 826–29.
5. *Time*, June 24, 1996, p. 62.
6. For previous controlled studies of intercessory prayer, see P. J. Collipp, "The Efficacy of Prayer: A Triple-blind Study," *Medical Times* 97 (1969): 201–204, and C. R. B. Joyce and R. M. C. Weldon, "The Objective Efficacy of Prayer: A Double-blind Clinical Trial," *Journal of Chronic Diseases* 18 (1965): 367–77.
7. Irwin Tessman and Jack Tessman, *Science* (April 18, 1988): 369–70.
8. For an account of the battle against emerging diseases, see

Laurie Garrett, *The Coming Plague: Newly Emerging Diseases in a World Out of Balance* (New York: PenguinUSA, 1994).

6. Close Encounters of the Improbable Kind

1. John Mack, *Abduction: Human Encounters with Aliens* (New York: Scribners, 1994). See also Keith Thompson, *Angels and Aliens: UFO's and the Mythic Imagination* (Reading, Mass.: Addison-Wesley, 1991) and David Jacob, *Secret Life: Firsthand Accounts of UFO Abductions* (New York: Simon & Schuster, 1992).
2. For a study of John Mack's subjects, see Joe Nickell, "A Study of Fantasy Proneness in the Thirteen Cases of Alleged Encounters in John Mack's Abduction," *Skeptical Inquirer*, May–June 1996, p. 18.
3. Frank J. Tipler, *The Physics of Immortality: Modern Cosmology, God, and the Resurrection of the Dead* (New York: Doubleday, 1994).

7. The New Story of Creation

1. Thomas Berry, "The New Story: Comments on the Origin, Identification and Transmission of Values," *Teilhard Studies* 1 (Winter 1978): 1.
2. Jacob Bronowski, *Science and Human Values* (revised ed.) (New York: Harper and Row, 1965), p. 119.
3. Joseph B. Soloveitchik, *The Lonely Man of Faith* (New York: Doubleday, 1996).
4. C. F. Kelley, *Meister Eckhart on Divine Knowledge* (New Haven: Yale University Press, 1977), p. 15.

8. Creationism and Evolution

1. John Paul II, "Message to the Pontifical Academy of Sciences on Evolution," *Origins*, November 14, 1996, p. 349.
2. For a report of Louis Sheldon's remarks, see *Boston Globe*, September 11, 1989, p. 9.
3. Harry Harlow, "Love in Infant Monkeys," *Scientific American*, July 1959, p. 68.
4. Jeremy B. C. Jackson and Alan H. Cheetham, *Paleobiology* 20 (1994): 407. For a summary and context for this article, see Richard A. Kerr, "Did Darwin Get It All Right?" *Science* 267 (March 10, 1995), p. 1,421.

5. Michael Pitman, *Adam and Evolution: A Scientific Critique of Neo-Darwinism* (Grand Rapids, Mich.: Baker Book House, 1984), p. 216.
6. I. L. Cohen, *Darwin Was Wrong* (Greenvale, N.Y.: New Research Publications, 1984), p. 116.
7. Dan Nilsson and Susanne Pelger, *Proceedings of the Royal Society* B256 (1994): 53–58.
8. Richard Dawkins, "The Eye in a Twinkling," *Nature* 368 (April 21, 1994): 690.
9. Jonathan Weiner, *The Beak of the Finch: A Story of Evolution in Our Time* (New York: Knopf, 1994).

9. Academy in Revolt

1. Bryan Appleyard, *Understanding the Present: Science and the Soul of Modern Man* (New York: Doubleday, 1993).
2. *Nature* 356 (April 30, 1992): 729.
3. For a report on Václav Havel's address before the World Economic Forum, see *New York Times*, March 1, 1992, p. E15.
4. E. R. Dodds, *The Greeks and the Irrational* (Boston: Beacon Press, 1957). See also Gerald Holton, *Science and the Irrational* (Boston: Harvard University Press, 1993), p. 149.
5. Paul R. Gross and Norman Levitt, *Higher Superstition: The Academic Left and Its Quarrels with Science* (Baltimore: Johns Hopkins University Press, 1994).
6. Alan Sokal, "Transgressing the Boundaries: Towards a Transformative Hermeneutics of Quantum Gravity," *Social Text*, Spring/Summer 1996, p. 217.
7. Alan Sokal, "A Physicist Experiments with Cultural Studies," *Lingua Franca*, May/June 1996, pp. 62–63.
8. Dai Rees, *Nature* 363 (May 20, 1993): 203.

10. In Search of the Self

1. E. O. Wilson, *Biophilia* (Boston: Harvard University Press, 1984), p. 46.
2. Gross and Levitt, *Higher Superstition: The Academic Left and Its Quarrels with Science*, p. 52.
3. Roger Penrose, *The Emperor's New Mind* (Oxford: Oxford University Press, 1989).

4. David Toolan, "Nature Is a Heraclitean Fire: Reflections on Cosmology in an Ecological Age," *Studies in the Spirituality of Jesuits*, November 1991, p. 31.
5. Mary Oliver, "Some Questions You Might Ask," from *New and Selected Poems* (Boston: Beacon Press, 1992), p. 65.

11. Knowing the Mind of God

1. Stephen Hawking, *A Brief History of Time* (New York: Bantam Books, 1988), p. 175.
2. Steven Weinberg, *Dreams of a Final Theory* (New York: Pantheon, 1992), p. 52.
3. E. O. Wilson, *Biophilia* (Boston: Harvard University Press, 1984), p. 10.
4. For discussion of accelerators and cathedrals, see Leon Lederman, with Dick Teresi, *The God Particle* (Boston: Houghton Mifflin, 1993), p. 254.
5. Nikos Kazantzakis, *The Saviors of God: Spiritual Exercises* (New York: Simon & Schuster, 1960), p. 101.

12. The Weight of Facts

1. *Science* 272 (May 10, 1996): 849.
2. Mary Oliver, "The Ponds," from *New and Selected Poems*, p. 93.

13. Work of the Eyes, Work of the Heart

1. Thomas Berry, "Future Forms of Religious Experience," *Riverdale Papers*, vol. 5, ed. by Thomas Berry (Riverdale, N.Y.: Riverdale Center), p. 2.
2. Mary Oliver, "In Blackwater Woods," from *New and Selected Poems*, p. 178.
3. William Butler Yeats, "He Wishes for the Cloths of Heaven," *The Collected Works of W. B. Yeats, Volume I: The Poems*, revised and edited by Richard J. Finneran (New York: Scribner, 1997), p. 73.

14. The Hallowing of the Everyday

1. Martin Buber, *I and Thou* (New York: Scribner, 1958).
2. Teilhard de Chardin, *The Phenomenon of Man* (New York: Harper, 1959).

Index